"十三五"职业教育国家规划教材

计算机专业英语

第五版

新世纪高职高专教材编审委员会 组编

主 编 卢川英 邵奎燕

副主编 来永春

大连理工大学出版社

图书在版编目（CIP）数据

计算机专业英语 / 卢川英，邵奎燕主编. — 5版
. — 大连：大连理工大学出版社，2018.2（2022.1重印）
新世纪高职高专计算机应用技术专业系列规划教材
ISBN 978-7-5685-1262-6

Ⅰ. ①计… Ⅱ. ①卢… ②邵… Ⅲ. ①电子计算机—
英语—高等职业教育—教材 Ⅳ. ①TP3

中国版本图书馆CIP数据核字(2018)第000077号

大连理工大学出版社出版

地址：大连市软件园路 80 号　邮政编码：116023

发行：0411-84708842　邮购：0411-84708943　传真：0411-84701466

E-mail: dutp@dutp.cn　URL: http://dutp.dlut.edu.cn

辽宁星海彩色印刷有限公司印刷　　大连理工大学出版社发行

幅面尺寸:185mm×260mm　印张：11.5　字数：264千字

2003年9月第1版　　2018年2月第5版

2022年1月第9次印刷

责任编辑：马　双　　责任校对：李　红

封面设计：张　莹

ISBN 978-7-5685-1262-6　　定价：37.50 元

前　言

《计算机专业英语》（第五版）是“十三五”职业教育国家规划教材、“十二五”职业教育国家规划教材、普通高等教育“十一五”国家级规划教材、高职高专计算机教指委优秀教材，也是新世纪高职高专教材编审委员会组编的计算机应用技术专业系列规划教材之一。

众所周知，英语在计算机领域及IT行业有着举足轻重的作用。新世纪信息技术的专业人才，不仅需要掌握扎实的专业基础知识和基本技能，还应当具备一定的英语运用能力，高职院校的学生更应注重英语运用能力的培养。因此，我们从高职院校人才培养要求出发，编写了这本独具特色的《计算机专业英语》（第五版）教材。

本教材特点如下：

1.本教材以学生在全英文环境下组装计算机并完成一个网上商店的设计为总任务，总任务下包括8个项目，通过各项目的理论知识和实践操作，使学生顺利地实现计算机的组装和网上商店的设计。

2.每个子项目采用结构化设计思路，分成3个部分，分别是理论学习、实践练习、职场英语。实践练习部分要求学生在全英语的环境下，在专业教师的指导下，在实验室完成相应的工作任务，力求使学生通过理论知识的学习，完成相应的计算机专业任务，并在完成专业任务的过程中巩固和检验理论知识的学习情况；针对计算机职业岗位需求，在每个单元中设计了Occupation English（职场英语）部分。模拟实际的工作情境进行未来工作岗位职场英语的对话，使学生在工作岗位中（或求职过程中）真正做到会听、会说、会用。

3.针对在科技英语阅读过程中普遍存在的“病症”，将每个单元分为几个训练模块，如句子主干训练、关键词训练、猜词训练等，每个训练模块训练目标明确，将普通英语中的阅读方法贯穿于专业英语的教学中，针对性地加以训练，使读者能够切实地掌握科技英语的阅读技巧，提高科技英语的阅读能力；文中生词的注释不像以往放在文章结束的

地方，而是标示在正文的一侧，方便读者阅读；每个单元的结尾还增加了构词法，使读者能够通过一词掌握多词，迅速扩充词汇量。

4.本教材内容取材新颖，多数取材来源于英文网站和英文资料的原版文章，力求反映计算机方面的新知识和新技术；任务设计紧密结合计算机专业类课程，难度适中。

全书共分为8个项目，分别为：计算机硬件的组装；计算机安装所需的软件；网络的组建与连接；网上商店的设计；网上商店的美化；商品数据库的建立；网上商店的安全问题；网上商店的交易。

本教材由吉林交通职业技术学院卢川英、邵奎燕任主编，石家庄工程职业学院来永春任副主编，宁波职业技术学院邱立中、吉林网格信息技术有限公司马贺参与编写。具体分工如下：项目1、3、7由卢川英编写；项目2、4由邵奎燕编写；项目5、6由来永春编写，Occupation English部分由邱立中编写，项目8由马贺编写。本教材由卢川英、邵奎燕统稿。

尽管我们在本教材的编写方面做了很多努力，但由于编者水平有限，加之时间紧迫，不当之处在所难免，恳请各位批评指正，以便下次修订时改进。

编 者

2018年2月

所有意见和建议请发往：dutpgz@163.com

欢迎访问职教数字化服务平台：http//sve.dutpbook.com

联系电话：0411-84707492　84706104

Contents

Navigation

我们在阅读科技英语的过程中，普遍存在若干“病症”，下面列举一些主要“病症”并辅以“对症良药”。

病症之一：科技英语中的句子有别于我们在其他外语书中常见到的句子，其从句、插入句、倒装句相对较多，不少初学者很不习惯，以致常常读不懂句子。

对症良药：尝试快速分辨句子的主、谓、宾，迅速读出句子的主干，掌握大意。

Passage One：主要训练分辨句子的主、谓、宾，把握句子主干，掌握句子以及段落的大意。主语用___表示，谓语用___表示，宾语用~~~表示，从句引导词用□表示。

病症之二：普通英语注重“精读”，即“word by word”，“sentence by sentence”。因而大多数读者在阅读科技英语时都是在逐字逐句（word by word）地读，极大地影响了阅读速度和阅读效果。

对症良药：专业英语注重的是“泛读”，抓住文章或段落的关键词（keyword），通过关键词来了解段落乃至句子的意思，围绕关键词再根据实际情况进行精读或慢读。

Passage Two：主要训练快速抓住关键词来掌握句子及段落的大意，提高阅读速度。

病症之三：我们从小到大都养成了遇到生词就查字典的习惯，一篇文章总有若干生词，几次查下来，阅读速度大打折扣，且极大地影响阅读兴趣。

对症良药：①培养猜词技能。遇到生词姑且放过，先不理会，继续往下读，然后根据前后文猜测出该词的意思。②学习词根构词法。英语中很多单词都是由词根衍变而来的，记住词根会大大提高背单词的效率。

Reading：主要训练猜词技巧。每个单元后的Word Building将帮助你快速掌握构词法。

病症之四：我们从小到大学了很多年的英语，但是真正到了工作岗位中（或求职过程中）却根本不会用。

对症良药：在英语学习的过程中注重多读、多听、多说。尤其是针对职业的岗位需求，模拟实际的工作情境进行未来工作岗位职场英语对话。

Occupation English：计算机专业学生未来工作岗位职场英语对话。

功夫不负有心人，相信读者掌握了正确的科技英语的学习方法，经过一段时间的训练后，您的专业英语的学习技能会有极大的提高。

Project One

Assembling Computers

Part A Theoretical Learning

Part B Practical Learning

Part C Occupation English

Part A Theoretical Learning

Training Target

In this part, our target is to improve the speed of reading professional articles and the comprehension ability of the reader. We have marked specialized vocabulary key words in some paragraphs so that the reader can quickly grasp the main idea of the sentences and paragraphs.

Skill One A Short Introduction to Computer

A computer is just a machine, but a computer system consists of two main elements: machine and **program**. Like a person, a computer system is composed of two parts: the bone—**hardware** and the soul—**software**. The central idea of a computer system is that input is processed into output. Input is the data which is entered into the computer, and output is the result of processing done by the computer, usually printed out or displayed on the screen.

Let us get closer to the computer from the basic **components**. When talking about computers, such image as Pic 1.1 will appear in our mind: a **display screen** known as the basic **output device**, a **keyboard** usually together with a **mouse** as the basic **input device**, and a **cabinet** known as a machine box.

program ['prəugræm]
n. 程序
hardware ['hɑ:dweə]
n. 硬件
software ['sɔftweə]
n. 软件

component [kəm'pəunənt]
n. 部件
display screen 显示器
output device 输出设备
keyboard ['ki:bɔ:d]
n. 键盘
mouse [maus] n. 鼠标
input device 输入设备
cabinet ['kæbinit]
n. 匣子，机箱

Pic 1.1 A computer's basic components

With the development of science and technology, the modern computer becomes more and more **flexible**, and the hardware family becomes stronger and stronger. A lot of new **peripherals** have appeared. These peripherals can be classified into two groups—input devices and output devices.

flexible ['fleksəbl]
adj. 灵活的
peripheral [pə'rifərəl]
n. 外围设备

Input devices (Pic 1.2), as the name suggests, are any hardware components that allow you to put the data, programs and commands into the computer. One of the most important input devices is the keyboard. Users can type in text or enter keyboard commands using the keyboard. Another device which can be used to input data is **scanner**. This electronic device is used to transfer an image such as text, or pictures into the computer. The most useful **pointing device** is a mouse, which allows the user to point to elements on the screen. There are some other input devices, such as **microphone, PC camera, digital camera, joystick, graphics tablet** and **light pen.**

scanner ['skænə]
n. 扫描仪
pointing device 定点设备
microphone ['maikrəfəun]
n. 麦克风
PC camera 电脑摄像头
digital camera 数码相机
joystick ['dʒɔi,stik]
n. 控制杆
graphics tablet 图形输入板
light pen 光笔

Pic1.2 Input devices

Output devices (Pic 1.3)are devices that let you see what the computer has accomplished. Several devices are used to display the output from a computer. **The favorite monitor is the LCD, which is slim and takes up little space, and displays text and images with greater clarity.** Another important output device is the printer,

LCD (Liquid Crystal Display)
abbr. 液晶显示器

which allows the user to copy the data in the computer onto the paper. **Speakers** and **headphones** allow the listener to hear **audio** data through the computer, such as speech or music. There are some other output devices, such as **projectors** and **facsimile machines**.

speaker ['spi:kə] n. 扬声器
headphone ['hedfəun] n. 耳机
audio['ɔ:diəu] n. 音频
projector [prə'dʒektə] n. 投影仪
facsimile machine 传真机

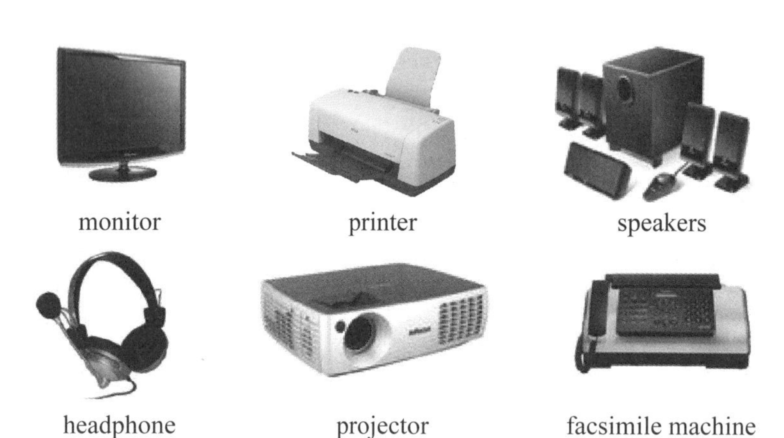

Pic 1.3 Output devices

All the components of a computer that we can see and feel are hardware. They work together to help us with our daily work.

Do you know how a computer can manage so many devices? The real secret lies in the machine box. When we take the cover off a small computer and look inside, the real computer appears in front of us, we will see a few circuit boards, some **wires** and some **cables**. In fact, the **motherboard** is the most important part in the machine box. Two main components on the motherboard are the **CPU** and **memory**.

wire ['waiə] n. 导线
cable ['keibl] n. 电缆
motherboard ['mʌðə:bɔd] n. 母板，主板
CPU (Central Processing Unit) abbr. 中央处理器
memory ['meməri] n. 内存
cook [kuk] n. 厨师

The CPU is sometimes referred to as the processor. It is the electronic device that interprets and carries out the basic instructions that operate the computer. The CPU is the control and data processing center of the whole computer system. You can simply regard it as a skillful **cook**, the only difference is just that the cook processes meat and vegetables, and makes them become delicious dishes. Here meat and vegetables are the input for the cook, and some dishes certainly are the output from the cook. Now turn back to our CPU, it can process the **digital data** from any input devices, and output them to an output device.

digital data 数字数据

Just like the excellent cook we mentioned before, he must need a number of blank plates around him, which stored meat and vegetables to be processed by the cook, that's a great help for his cooking. And after dinner, the plates should be cleaned up. Memory stores information processed by the CPU. The **data stream** can flow from

data stream 数据流

the CPU into memory or on the contrary. Memory consists of **RAM** and **ROM**. Any information in RAM will be lost when the computer is turned off, just like the plates that are cleaned by the cook.

Most of the devices connected to the computer communicate with CPU in order to carry out a task. The CPU controls the data flow on the **inner Bus**, there are three kinds of Buses used in our computer: **AB, DB** and **CB**. The most popular Bus to be used on a motherboard is a **PCI** Bus which is a peripheral component interface Bus.

The CPU uses storage to hold data, instructions and information for future use. Storage is also called secondary storage or **auxiliary storage**. Think of storage as a little cabinet used to hold **file folders**, and memory as the top of your desk. When you need a file, you can get it from the **filing cabinet** (storage) and place it on your desk (memory). When you finish a file, you return it to the filing cabinet (storage).The items in storage are **retained** even when power is removed from the computer.

.End.

RAM (Random Access Memory) abbr. 随机存取存储器
contrary ['kɔntrəri] n. 相反
ROM (Read Only Memory) abbr. 只读存储器
inner Bus 内部总线
AB (Address Bus) abbr. 地址总线
DB (Data Bus) abbr. 数据总线
CB (Control Bus) abbr. 控制总线
PCI(Peripheral Component Interconnect) abbr. 外设部件互连
auxiliary storage 辅助存储器
file folder 文件夹
filing cabinet 文件柜
retain [ri'tein] v. 保留

Key Words

hardware n. 硬件
keyboard n. 键盘
printer n. 打印机
wire n. 导线
motherboard n. 母板，主板
headphone n. 耳机
audio n. 音频
joystick n. 控制杆
peripheral n. 外围设备
input device 输入设备
light pen 光笔
inner Bus 内部总线
facsimile machine 传真机
data stream 数据流
PC camera 电脑摄像头
graphics tablet 图形输入板
software n. 软件
mouse n. 鼠标
projector n. 投影仪
cable n. 电缆
memory n. 内存
speaker n. 扬声器
scanner n. 扫描仪
microphone n. 麦克风
output device 输出设备
display screen 显示器
auxiliary storage 辅助存储器
file folder 文件夹
digital data 数字数据
pointing device 定点设备
digital camera 数码相机

CPU (Central Processing Unit) abbr.中央处理器
RAM (Random Access Memory) abbr.随机存取存储器
ROM (Read Only Memory) abbr.只读存储器
PCI (Peripheral Component Interconnect) abbr.外设部件互连
AB (Address Bus) abbr.地址总线
DB (Data Bus) abbr.数据总线
CB (Control Bus) abbr.控制总线

参考译文 技能1 计算机简介

一台计算机只是一部机器，而一个计算机系统则包括两个要素：机器和程序。像人一样，计算机系统也由两部分组成：硬件（像人的骨架）和软件（像人的灵魂）。计算机系统的核心思想是把输入处理成为输出。输入是进入计算机的数据，而输出则是计算机处理的结果，一般由打印机打印或显示器显示出来。

我们就计算机的基本构成来认识计算机。当我们谈论计算机时，会在脑海中出现如图1.1所示的画面。显示器是基本的输出设备，键盘和鼠标一起作为基本的输入设备，还有一个称为机箱的盒子。

随着科学技术的发展，现代计算机也越来越便捷，硬件家族也越来越强大。出现了许多新式外围设备，这些外围设备可以分为两类——输入设备和输出设备。

顾名思义，输入设备(图1.2)是让用户向计算机输入数据、程序和命令的硬件部件。最重要的输入设备之一是键盘。用户可以用它录入文本或输入键盘命令。可用来输入数据的另一种设备是扫描仪。这种电子设备可将文本、图片等影像传入计算机。最有用的定点设备是鼠标，它使用户能够指向屏幕上的内容。还有一些其他的输入设备，如麦克风、电脑摄像头、数码相机、控制杆、图形输入板和光笔。

输出设备（图1.3）可以让你看到电脑已经完成的内容。有几种设备可以用于计算机的输出显示。比较流行的显示器是液晶显示器，它体积小，占地少，能更清晰地显示文本和图像。另一种重要的输出设备是打印机，用户能用它将计算机中的数据拷贝到纸上。扬声器和耳机能使用户通过计算机听到演讲或音乐等音频数据。还有一些其他的输出设备，如投影仪和传真机。

所有我们看得见摸得着的部件都叫作硬件。它们相互协作帮助我们处理日常工作。

你知道计算机怎样管理这么多的设备吗？真正的秘密就在机箱里，当我们把小型的计算机的机箱盖取下并观察其内部的时候，一台真正的计算机就展现在我们面前了。我们会看到几块电路板、一些导线和一些电缆。事实上，在机箱里最重要的部件是主板，在主板上有两个主要的部件：中央处理器和内存。

中央处理器有时也被称为处理器，该电子设备用来解释并执行计算机的基本操作指令。中央处理器是整个计算机的系统控制和数据处理中心。你可以简单地将其当作一个熟练的厨师，不同之处只不过是厨师加工肉类和蔬菜，把它们变成美味的佳肴。这里肉类和蔬菜是输入给厨师的，而这些佳肴当然是从厨师那里输出的。现在回到我们的中央处理器，它可以处理从输入设备输入的数字数据，并将其输出到某种输出设备上。

正如我们刚才提到的那个优秀的厨师，在他周围一定需要许多空盘子，这些盘子用于存储他将要处理的肉类和蔬菜，对他做菜有很大的帮助，而且在宴会之后，盘子就会被清洗干净。内存存储中央处理器所处理的信息。数据流可以从中央处理器流入内存或者从内存流入中央处理器。内存由随机存取存储器和只读存储器组成。当计算机关机时，在随机存取存储器中的信息将会丢失，正如同盘子被厨师清洗干净一样。

大多数连接到计算机上的设备通过与中央处理器通信来完成作业。中央处理器控制着内部总线上的数据流，计算机有三种总线：地址总线、数据总线和控制总线。在主板上最常用的总线是PCI总线，即外设部件互连总线。

中央处理器采用存储器保存将来要使用的数据、指令和信息。存储器也叫第二存储器或辅助存储器，你可以把存储器看作是一个用来存放文件夹的小柜子，把内存当作办公桌面。当你需要一份文件时，你可以从文件柜（存储器）中获取并把它放置在办公桌面（内存）上，当你用完这份文件后，你又可以把它放回文件柜（存储器）。存储器里的内容即使是在计算机断电时也能被保存。

Skill Two How Does a Computer Work?

When we talk about the computer, we usually meet the topic: motherboard (Pic 1.4). The main circuit board in a computer is called the motherboard. It is a flat board that holds all of the key elements that make up the “brain” of the system, including the **microprocessor** or CPU, RAM or primary memory, and expansion slots which are **sockets** where other circuit boards called expansion boards may be **plugged in**.

microprocessor [maikrəu'prəusesə] n. 微处理器
socket ['sɔkit] n. 插座
plug in 插入

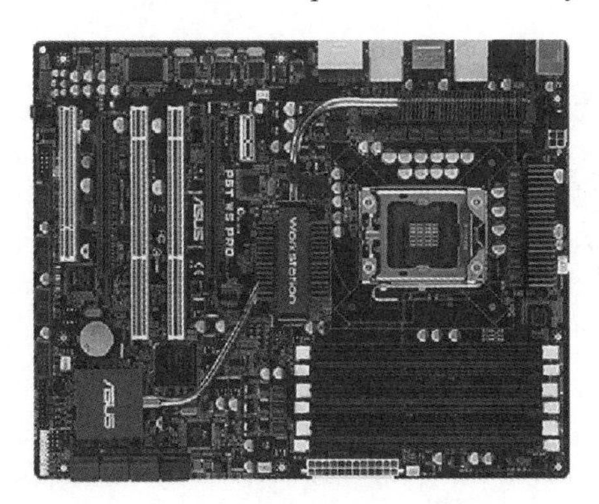

Pic 1.4 Motherboard

Let us use the system presented in Pic 1.5 to show you how a typical computer works. A computer is controlled by a stored program, so if we want to use a computer, the first step is to copy the program from diskette into memory. Now the **processor** can begin **executing** instructions; the data input from the keyboard is stored in memory. The processor processes the data and then stores the results back into memory. At last, we can get the result.

processor ['prəusesə]
n. 处理器
execute ['eksikju:t]
v. 执行

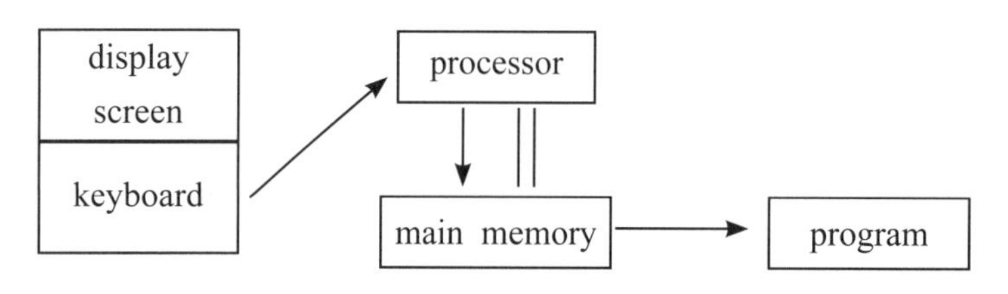

Pic 1.5 Computer system

Now we can see that a computer system consists of four basic components. An input device provides data. The data is stored in memory, which also holds a program. Under the control of the program, the computer's processor processes the data. The results flow from the computer to an output device. Let us introduce the system components one by one, beginning with the processor.

The processor, usually called the Central Processing Unit (CPU, Pic 1.6) or main processor, is the heart of a computer. It is the CPU that in fact processes or **manipulates** data and controls the rest part of the computer. How can it manage its job? The secret is software. Without a program to provide control, a CPU can do nothing. How can a program guide the CPU through the processes? Let us consider from the basic element of a program—instruction. An instruction is composed of two parts: an **operation code** and one or more **operands** (Pic 1.7). The operation code tells the CPU what to do and the operands tell the CPU where to find the data to be manipulated.

manipulate [mə'nipjuleit]
vt. 处理，操作

operation code 操作码
operand ['ɔpərænd]
n. 操作数

Pic 1.6 CPU

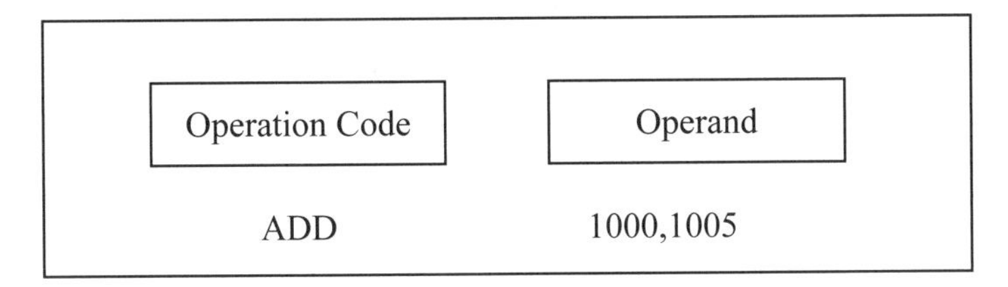

Pic 1.7 Instruction

The processor contains four major components (Pic 1.8): a clock, an instruction control unit, an arithmetic and logic unit (usually shortened to ALU) and several **registers**. The clock generates **precisely** timed **pulses** of **current** that **synchronizes** the processor's other components. Then the instruction control unit determines the location of the next instruction to be executed and fetches it from the main memory. The arithmetic and logic unit performs arithmetic operations (such as addition and subtraction) and logic operations (such as testing a value to see if it is true), while the registers are **temporary** storage devices that hold control information, key data and some **intermediate** results. Since the registers are located in the CPU, the processing speed is faster than the main memory. Then which is the key component to a computer's speed? It is the clock! In more detail, it is the clock's **frequency** that decides a computer's processing speed. When we buy a computer, we usually consider the main frequency first, and that means a clock's frequency.

register ['redʒistə] n. 寄存器
precisely [pri'saisli] adv. 精确地
pulse [pʌls] n. 脉冲
current ['kʌrənt] n. 电流
synchronize ['siŋkrənaiz] v. 同步
temporary ['tempərəri] adj. 暂时的
intermediate [ˌintə'mi:diət] adj. 中间的
frequency ['fri:kwənsi] n. 频率

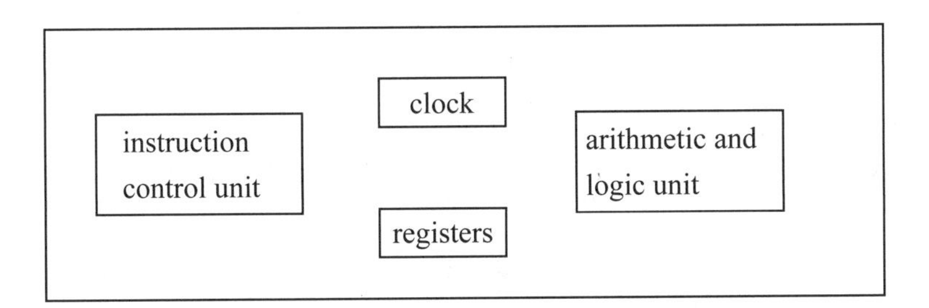

Pic 1.8 Processor's four major components

Now we will talk more in detail about the Microprocessors and Central Processing Units. Microprocessors are central processing units etched on a tiny chip of silicon and, thus, are called microchips. Microprocessors contain many electronic **switches**, called transistors, which determine whether electric current is allowed to pass through. Transistors are the basic building blocks of microprocessors. A single microchip may contain millions of transistors. When electric current is allowed to pass through, the switch is on. This **represents** a 1 bit.

switch [switʃ] n. 开关
represent [repri'zent] v. 代表

If the current is not allowed to pass through, the switch is off. This represents a 0 bit. Different **combinations** of transistors represent different combinations of bits, which are used to represent special characters, letters, and digits.

combination [kɔmbi'neiʃən] n. 组合

.End.

Key Words

microprocessor n. 微处理器	processor n. 处理器
operand n. 操作数	register n. 寄存器
frequency n. 频率	switch n. 开关
socket n. 插座	pulse n. 脉冲
synchronize v. 同步	plug in 插入
operation code 操作码	

参考译文 技能2 计算机是如何工作的?

当我们谈及计算机的时候，我们通常会遇到这样一个话题：主板（图1.4）。计算机内的主要电路板叫主板。它是一个平板，存放着所有组成系统“大脑”的关键元素，包括微处理器（或CPU）、RAM（或主存），还有一些扩展槽，它们是一些插口，可以将其他电路板（扩展板）插到里面。

让我们用图1.5所示的系统来说明一个典型的计算机是如何工作的。计算机是受存储程序控制的，所以，如果想使用计算机，第一步要把程序从磁盘上拷贝到主存中。现在，处理器能开始执行指令了，从键盘输入的数据也存在主存中。用处理器加工数据，然后把结果存回主存。最后，我们就得到了结果。

现在我们可以看到计算机由四个基本的部件组成：提供数据的输入设备、存放数据的存储器（程序也保存在内）、在程序的控制下处理数据的处理器、输出结果的输出设备。让我们逐个介绍这些部件，先从处理器开始。

处理器通常称为中央处理器（CPU）或主处理器（图1.6），是计算机的心脏。事实上，是由CPU来处理或操作数据，并且控制计算机的其他部件的。它是怎样完成工作的呢？答案是软件。如果没有程序提供控制，CPU什么都不能做。程序怎样全程引导CPU工作呢？让我们从程序的基本元素——指令开始讲述。一条指令由两部分组成：一个操作码和一个或多个操作数（图1.7）。操作码告诉CPU做什么，操作数告诉CPU到哪儿去找要被操作的数据。

处理器包含四大部件（图1.8）：时钟、指令控制单元、算术逻辑运算单元（简称ALU）和一组寄存器。时钟始终精确地产生定时电流脉冲，使之与处理器的其他部件同步。然后指令控制单元确定下一条指令的地址，并从主存中取出指令。算术逻辑运算单元执行算数操作（例如加和减）和逻辑操作（例如测试数值的真假）。寄存器是临时存储器件，它保存的是控制信息、关键数据和一些中间结果。因为寄存器在CPU上，所以它的处理速度比主存储器快。那么决定计算机速度的关键部件是什么呢？时钟！详细地说，是时钟的频率决定了计算机的处理速度。当我们购买一台计算机的时候，首先要考虑主频，即时钟的频率。

现在我们将详细地谈一谈微处理器和中央处理器。微处理器是蚀刻在微小的硅质芯片上的中央处理单元，因此也叫微芯片。微处理器包含许多电子开关，叫做晶体管，它决定了电流是否可以通过。晶体管是微处理器的基本组成块。一个简单的芯片可包含数万个晶体管。当电流允许通过的时候，开关打开，这代表二进制位1。如果电流不允许通过，开关关闭，则代表二进制位0。不同的晶体管组合代表不同的二进制组合，可以用它来表示特定的字符、字母和数字。

Fast Reading One Input and Output

We often mention input/output system (or I/O), what's I/O system? In computing, I/O is the communication between an information processing system (such as a computer) and the outside world. Inputs are the signals or data received by the system, and outputs are the signals or data sent from it. I/O devices are used by a person (or other system) to communicate with the computer. For example, the keyboard and the mouse (Pic 1.10) may be an input device for the computer, while monitors (Pic 1.9) and printers are considered as output devices. Modem (Pic 1.11) and Network Interface Cards (NIC, Pic 1.12), typically serve for both input and output devices.

Pic 1.9 Monitor

Pic 1.10 Keyboard and mouse

Pic 1.11 Modem

Pic 1.12 NIC

The mouse and the keyboard are input physical devices. Users use them to input information, and then input devices convert it into the signal that a computer can understand. The output information from these devices is input for the computer. Similarly, the printer and the monitor take it as input signal that a computer outputs. They convert these signals into symbols that the user can see or read. These interactions between the computer and the user are called human-computer interactions.

Memory is the device that the CPU can read and write to directly, with individual instructions. In computer architecture, the combination of the CPU and the main memory is considered as the brain of a computer, and from that point of view, any transfer of information from or to that combination, for example to or from a disk drive, is considered as I/O. The CPU and its supporting circuitry provide memory-mapped I/O that is used in low-level computer programming, such as the implementation of device drivers. An I/O algorithm is one designed to exploit locality and perform efficiently when data resides on secondary storage, such as a disk drive.

A computer uses memory-mapped I/O accesses hardware by reading and writing to specific memory locations, using the same assembly language instructions that computer would normally use to access memory.

.End.

参考译文 输入/输出设备

我们常常说输入/输出系统（I/O），什么是输入/输出系统呢？在计算机里面，输入/输出系统是内部信息处理系统和外面世界之间的交流设备。输入是系统接收到的信号或是数据，输出是从系统传输出去的信号或数据，是人（或其他系统）用来和计算机进行交流的设备。例如，键盘和鼠标（图1.10）对于计算机来说是输入设备，而显示器（图1.9）和打印机通常被认为是输出设备。调制解调器（图1.11）和网卡（图1.12）既是输入设备又是输出设备。

鼠标和键盘作为物理的输入设备，用户使用它们输入信息，之后输入设备把这些信息转换成计算机能识别的信号。从这些设备输出的信息是计算机的输入信息。同样地，打印机和显示器把计算机的输出信号作为输入信号。它们把这些信号转换成用户可以明白或是可读的信息。这种计算机和用户之间的交流被称为人机交互。

主存是有独立指令的设备，CPU能直接进行读写。在计算机的架构体系中，CPU和主存的组合被认为是计算机的大脑，从这个观点出发，任何从大脑转换出的信息或到大脑的信息，例如，从大脑到磁盘驱动器或从磁盘驱动器到大脑，可认为是输入/输出（I/O）。CPU及其支持电路提供内存映射I/O，可用于低级计算机编程，如设备驱动程序的实现。一个I/O算法要在数据驻留在二级存储设备（如磁盘驱动器）时仍可有效利用位置并执行指令。

使用内存映射I/O访问的计算机硬件，可阅读和写作具体的内存位置，并使用计算机通常会使用的汇编语言指令来访问内存。

Fast Reading Two History of the ENIAC

The Electronic computer was one of the greatest inventions in the 20th century. Once talking about computers, we have to think of the birth of ENIAC(Electric Numerical Integrator And Calculator)(Pic 1.13).

Pic 1.13 ENIAC

The start of World War Ⅱ produced a large need for computer capacity, especially for the military. New weapons were made for trajectory tables and other essential data were needed. In 1946, John P. Eckert, John W. Mauchly, and their associates at the Moore School of Electrical Engineering at University of Pennsylvania decided to build a high-speed electronic computer to do the job. This machine became known as ENIAC.

The size of ENIAC's numerical "word" was 10 decimal digits, and it could multiply two of these numbers at a rate of 300 per second, by finding the value of each product from a multiplication table stored in its memory. ENIAC was therefore about 1,000 times faster than the previous generation of relay computers.

ENIAC used 18,000 vacuum tubes, about 1,800 square feet of floor space, weighed 30 tons and consumed about 180,000 watts of electrical power. It had punched card I/O, 1 multiplier, 1 divider/square rooter, and 20 adders using decimal ring counters, which served as adders and also as quick-access (0.0002 seconds) read-write register storage. The executable instructions making up a program were embodied in the separate "units" of ENIAC, which were plugged together to form a "route" for the flow of information.

ENIAC was commonly accepted as the first successful high-speed electronic digital computer (EDC) and had been used from 1946 to 1955, but it had a number of shortcomings which were not solved, notably the inability to store a program. A number of improvements were also made to ENIAC from 1948, based on the ideas of the Hungarian-American mathematician, John Von Neumann (Pic 1.14)who was called the father of computer.

Von Neumann contributed a new awareness of how practical, yet fast computers should be organized and built. These ideas, usually referred to as the stored-program technique, became essential for future generations of high-speed digital computers and were universally adopted. Electronic Discrete Variable Automatic Computer (EDVAC) designed by Von Neumann was built in 1952. This computer used 2,300 vacuum tubes, but its speed was 10 times faster than ENIAC which used 18,000 vacuum tubes. And the most important, Random Access Memory (RAM) was used.

Pic 1.14 John Von Neumann

.End.

参考译文 ENIAC的简介

电子计算机是20世纪最伟大的发明之一。一提起计算机，我们就不得不提到ENIAC（电子数字积分器和计算器）（图1.1）的问世。

第二次世界大战的爆发对计算机的能力提出了更高的要求，尤其是在军事领域上。新武器的制造需要弹道表和其他关键数据。1946年，John P. Eckert、John W. Mauchly和他们在宾州大学摩尔电器工程学院的同事决定制造一台高速电子计算机来完成这项工作。这台机器被称为ENIAC（埃尼亚克）。

ENIAC的数字“字长”为10位十进制数字，它以每秒300次的速度运算两个这样数字的乘法，其方法是从存储在它的存储器中的乘法表中找到每次乘积的值。因此，ENIAC具有比它的前一代继电器计算机快1000倍的速度。

ENIAC使用了18000个电子管，占地约1800平方英尺，重达30吨，消耗大约180000W电能，它有穿孔卡片I/O（输入/输出设备）、一个乘法器、一个除法器/平方根器和使用十进制循环计数器的20个加法器。这些设备既可用作加法器，又可快速访问（0.0002秒）读/写寄存存储器。在ENIAC里，可执行的指令组成一个程序，包含于单独的“单元”里，这些单元连接起来为信息流形成一个“路由”。

ENIAC被普遍认为是第一个成功的高速电子数字计算机，并在1946年至1955年中得到应用。但是它有许多缺点没有得到解决，尤其是不能存储程序。从1948年起，ENIAC有了许多改进，其思想来源于被人们称为“计算机之父”的美籍匈牙利数学家约翰·冯·诺依曼（图1.14）。

诺依曼提出了如何组建一个应用型的、快速运算的计算机的新见解。这些思想（通常指存储技术）对以后高速数字计算机的发展十分必要，因此被普遍采纳。1952年，由计算机之父冯·诺伊曼设计的电子计算机EDVAC问世。这台计算机总共使用了2300个电子管，运算速度却比拥有18000个电子管的“埃里亚克”提高了10倍。更重要的是，随机存取存储器（RAM）被采用了。

Ex 1 **What is a computer like in your mind? Try to give a brief summary of this passage in no more than five sentences.**

Ex 2 **Fill in the table below by matching the corresponding Chinese or English equivalents.**

display screen	
	打印机
scanner	
	内存
output device	
	主板
LCD	
	随机存取存储器
digital camera	
	文件夹
pointing device	
	传真机

Ex 3 **Choose the best answer to the following questions according to the text we learnt.**

1. A computer system consists of two main elements: ________.
   A. input devices, output devices
   B. hardware, software
   C. CPU, memory
   D. RAM, ROM
2. The peripherals of the computer can be classified into two groups: ________.
   A. data Bus, cables
   B. pointing device, output devices
   C. input devices, output devices
   D. control Bus, inner Bus
3. Users can type in text and commands using the ________, which is one of the most important input devices.
   A. mouse　　B. scanner
   C. microphone　　D. keyboard

4. The most useful pointing device is a ________, which allows the user to point to elements on the screen.

A. digital camera　　B. mouse

C. PC camera　　D. keyboard

5. The characteristic of the LCD is ________.

A. taking up a lot of space

B. taking up little space

C. displaying text and images with greater unclarity

D. A and C

6. ________is the control and data processing center of the whole computer system.

A. Memory　　B. Motherboard

C. CPU　　D. Bus

7. When the computer is turned off, any information in ________will be lost.

A. RAM　　B. ROM

C. storage　　D. memory

8. ________ is the most popular Bus to be used on a motherboard.

A. Address Bus　　B. Data Bus

C. Control Bus　　D. PCI Bus

9. Two main components on the motherboard are ________.

A. memory and circuit boards

B. CPU and memory

C. arithmetic unit and control unit

D. ROM and RAM

10. ________ does not belong to the output devices.

A. facsimile machine　　B. headset

C. speaker　　D. microphone

Part B Practical Learning

Training Target

In this part, students must finish two special tasks in English environment, under the guidance of the Specialized English teacher and the teaching related to the computer assembly and maintenance in the computer laboratory. First the teacher must divide the students into several groups.

Task One Name the Computer Hardware Devices

The first task is to name the computer hardware devices. In this task, students must know the composition of the computer hardware system (Pic 1.15 and Pic1.16), and their English names.

There is some information about the hardware devices of a computer. The information can help students finish the task.

As we know, the computer hardware system usually consists of Calculator, Controller, Memory and I/O . Calculator and Controller are often referred to CPU. Memory includes the main memory and the secondary storage (suck as hard disk). The basic input devices include keyboard and mouse, the basic output devices include display screen, printer and so on. These devices must plug in the motherboard through the sockets or slots (Pic1.17).

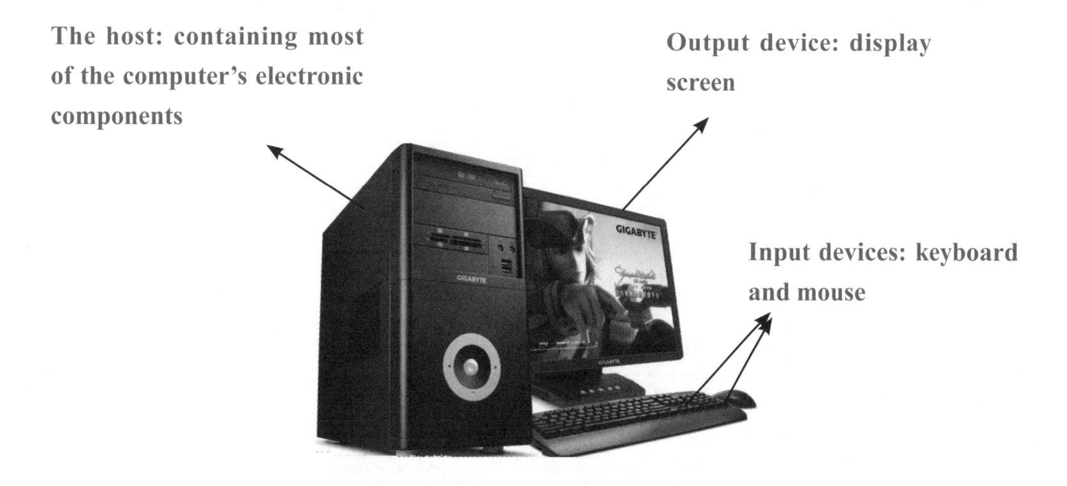

Pic 1.15

Pic 1.16

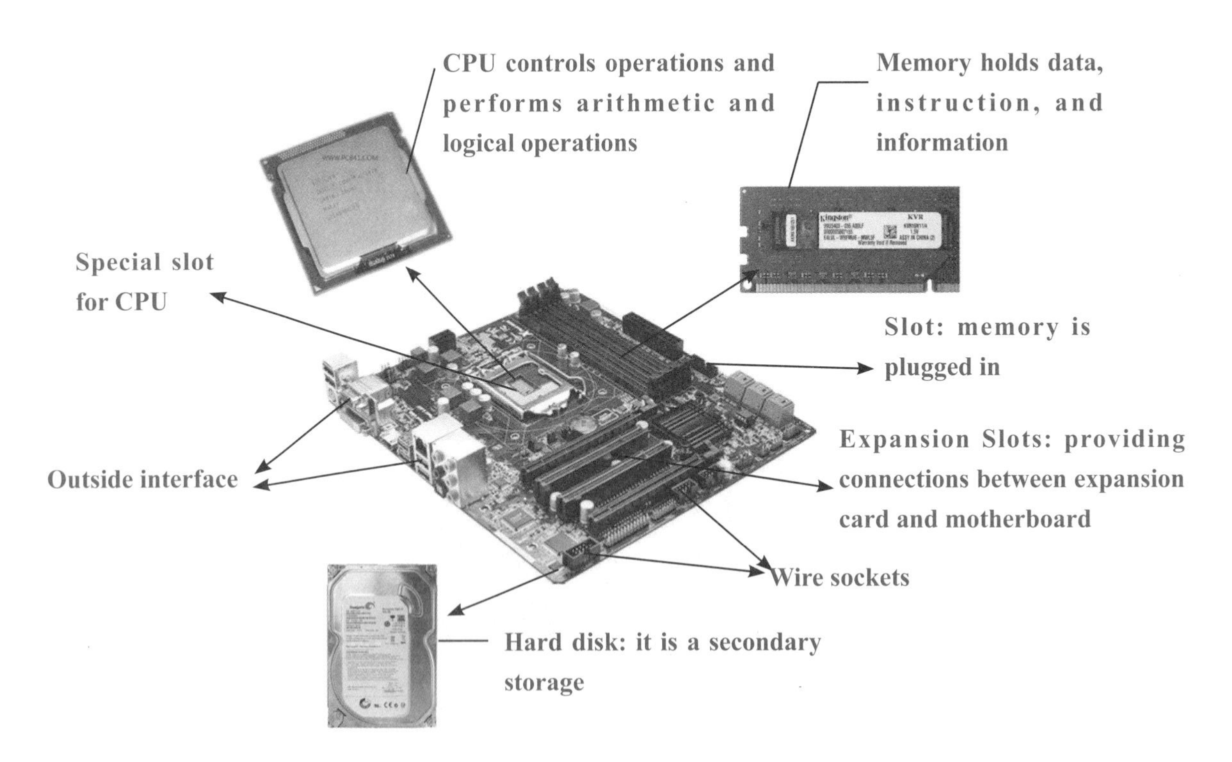

Pic 1.17

Task Two Computer Assembly and CMOS Setup

In this task, first students can assemble the computer. Secondly, students can set up the CMOS after finishing the assembling.

There is some information about the CMOS.

Complementary metal-oxide-semiconductor (CMOS) on the motherboard records the date, time, hard disk references, and other advanced references about computer. Knowing how to access and change settings in your BIOS or CMOS can save you a lot of headache when troubleshooting a computer.

First: How to enter the CMOS?

There are numerous ways to enter the CMOS setup. Below is a listing of the majority of these methods as well as other recommendations for entering the CMOS setup.

As the manufacturer's logo appears, press the designated setup button to enter the CMOS. The key is decided by the manufacturer. Typical setup keys are F2, F10, F12, ESC and DEL. The key will be displayed on the screen with the manufacturer's logo (Pic1.18).

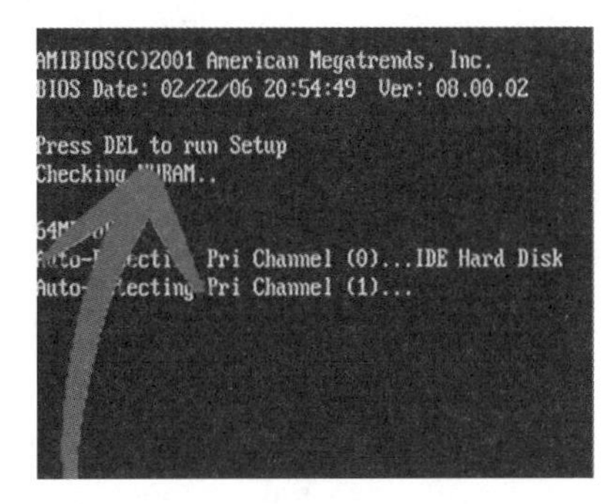

Pic 1.18

IF you strike the DEL key immediately, you will enter the CMOS setup (Pic 1.19).

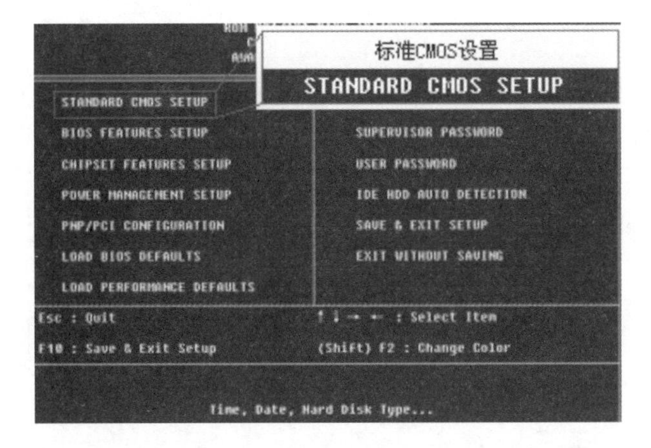

Pic 1.19

The details of CMOS setup (Pic 1.20).

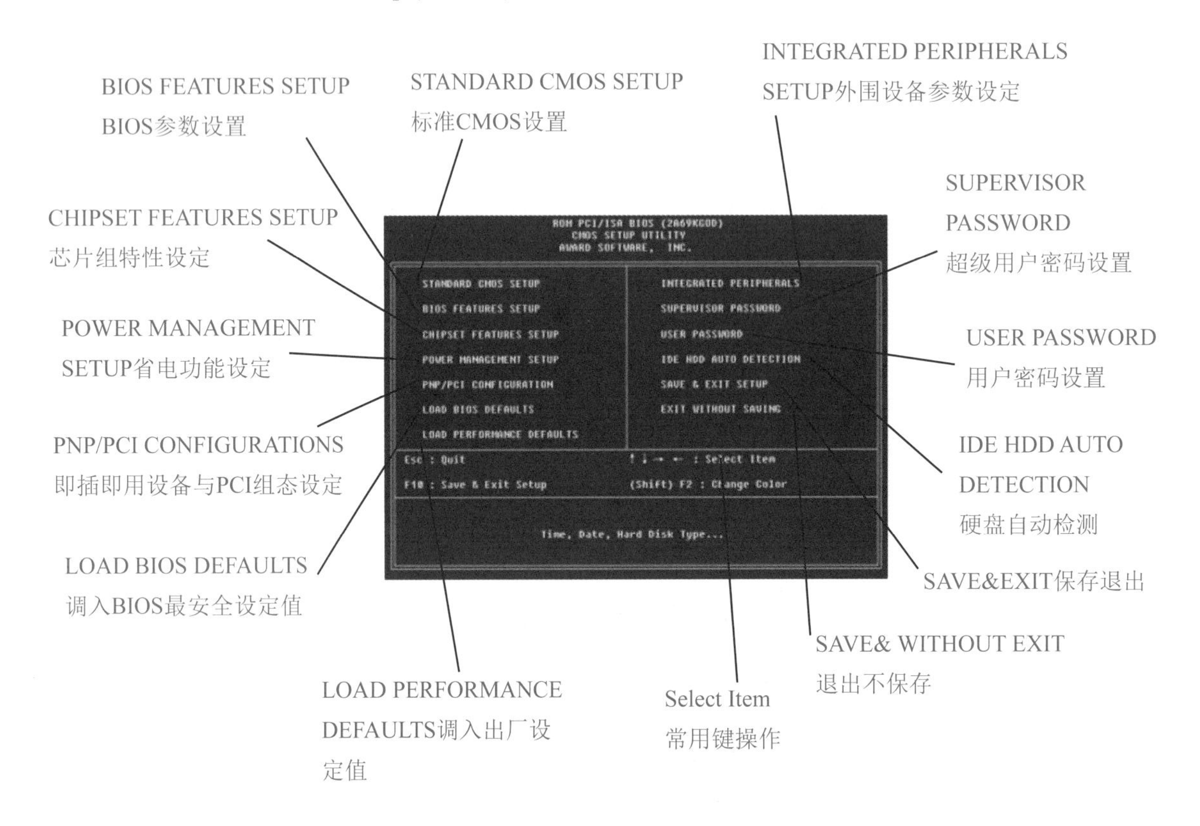

Pic1.20

Part C Occupation English

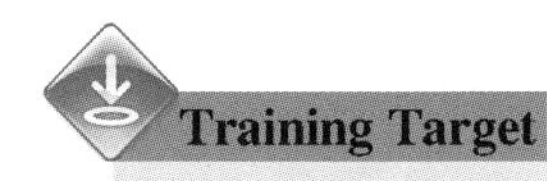

Training Target

In this part, students are supposed to practice the dialogue, which may occur every day in sale position. It is a basic ability to recommend the products on sale. Moreover, salespersons may try their best to make the transaction.

Introduction to the Computers on Sale
针对顾客购买计算机所做的介绍
Post: PC Salesperson（岗位：计算机销售）

A : Hello, welcome to Lenovo. Can I help you? 你好，联想笔记本电脑店欢迎你！需要什么服务吗？

B : Hello, I'd like to get some information about IdeaPad notebook. 你好！我想了解一下IdeaPad笔记本的有关信息。

A : Certainly, we have laptops for home, home office and everywhere in between. IdeaPad notebooks feature up-to-the-minute technology and cool features. They are designed for home finance, surfing the web, email, gaming, videos, presentations and so on. Further more, they are sleek, ultraportable and stylish. 好的，我们的笔记本电脑，无论居家、家庭办公或其他任何场合都可以使用。IdeaPad笔记本采用最新技术，具有完美品质。可用于家庭理财、网上冲浪、收发电子邮件、游戏、视频、授课等。而且，手感光滑、轻便易带、高雅华贵。

B : Can you tell me their features? 有哪些特色呢？

A : Sure, we use Intel® processor technology with Intel® Core 2 Duo processors and advanced wireless technology, offering notebooks with the new Windows Vista® operating system for added features and security. With Lenovo, you may enjoy advanced multimedia options such as an integrated camera, Blu-ray Disc (BD) drive, multimedia card reader, VGA, DisplayPort, and more.我们使用英特尔处理技术，英特尔双核处理器以及高级无线技术，提供最新的Windows Vista操作系统增加其功能性和安全性。联想电脑可以使你享受可供选择的高级多媒体技术，如一体化摄像、蓝光光碟驱动、多媒体读卡、VGA、显示接口技术等等。

The solid-state storage drives use flash memory rather than traditional magnetic storage. They are lighter than traditional drives, use less power and offer improved performance. Since they have no moving parts, solid-state drives are considered less susceptible to damage or breakdown. 固态硬盘使用闪存技术，淘汰了传统的磁储存，更加轻便，耗电更少，并提升了性能。由于没有可移动的部分，故固态硬盘不易损坏或出故障。

Besides, you can recover from a virus with press of a button. 除此以外，如果感染了病毒，还可以一键恢复。

The ThinkVantage Active Protection System uses an accelerometer to monitor movement of the system and stop the drive to help protect against damage when a fall or similar event is detected. 笔记本硬盘动态保护系统使用加速计来监控系统的运行，可帮助保护你的硬盘驱动器免受由强烈的物理震动导致的损坏。

B : Yes, I see. Thank you. 谢谢。

前缀/后缀由一个或几个字母组成，放在词根或单词之前/之后，组成一个新词。

（1）auto-（前缀）：自动的

alarm：报警器————autoalarm：自动报警器

code：编码————autocode：自动编码

（2）co-（前缀）：共同

action：行动————coaction：共同行动

operate：操作————cooperate：协作，合作

（3）fore-（前缀）：预……，前……

head：头 ————forehead：前额

word：话 ————foreword：前言，序言

（4）-ess（后缀）：女性的

mayor：市长————mayoress：女市长

actor：男演员————actress：女演员

（5）-ee（后缀）：被……的人，受……的人

employ：雇用 ————employee：雇员

test：测试 ————testee：被测验者

Ex Translate the following words and try your best to guess the meaning of the word on the right according to the clues given on the left.

correction 校正（名词） autocorrection ________________

biography 传记（名词） autobiography ________________

exist	存在（动词）	coexist ________
author	作家（名词）	coauthor ________
tell	告诉（动词）	foretell ________
father	父亲（名词）	forefather ________
waiter	男服务员（名词）	waitress ________
host	主人（名词）	hostess ________
train	训练（动词）	trainee ________
pay	薪水（名词）	payee ________

Exercise

Ex 1 **When we use a computer, what does the processor do?**

Ex 2 **Fill in the table below by matching the corresponding Chinese or English equivalents.**

processor	
	插座
microprocessor	
	寄存器
operation code	
	开关
main frequency	
	脉冲
operand	

Ex 3 **Choose the best answer to the following questions according to the text we learnt.**

1. Which of the following is not one of the four parts that comprise a processor? ________

   A. Clock. B. Memory.

   C. Register. D. ALU.

2. A computer system contains input device, output device, CPU and ________.

   A. ROM B. RAM

   C. memory D. register

3. The processor consists of ________.

   A. clock , ALU, register and operation code

   B. ALU, instruction control, clock and register

   C. CPU, memory, register and clock

   D. instruction control, clock, operation code and operand

4. ________ is the soul and heart of a computer, and it can manipulate data and control the rest part of the computer.

   A. Memory　　B. Input device

   C. Register　　D. CPU

5. Which of the following tells the CPU what to do? ________

   A. Instruction.　　B. Operand.

   C. Operation code.　　D. Arithmetic and logic unit.

6. Which of the following tells the CPU where to find the data to be manipulated? ________

   A. Clock.　　B. Instruction control unit.

   C. Operation code.　　D. Operand.

7. When we buy a computer, we usually consider the ________ first, and that means a clock's frequency.

   A. main frequency　　B. board

   C. display　　D. keyboard

8. The arithmetic unit performs the ________operations.

   A. arithmetic　　B. comparison

   C. logical　　D. all of the above

9. ________ holds control information, key data and some intermediate results.

   A. Register　　B. Instruction control unit

   C. ALU　　D. Storage

10. When electric current is not allowed to pass through transistors on the microchip, the switch is________ and it represents a ________ bit.

   A. off, 1　　B. on, 1

   C. on, 0　　D. off, 0

Project Two

Installing the Software

Part A Theoretical Learning

Part B Practical Learning

Part C Occupation English

Part A Theoretical Learning

Training Target

In this part, our target is to improve the speed of reading professional articles and the comprehension ability of the reader. We have marked specialized vocabulary keywords in some paragraphs, and also made the flexible sentences strong black and marked the subject, predicate and object of them. Try to grasp the main idea of these sentences. You can quickly grasp the main idea of the sentences and paragraphs.

Skill One System Software—OS

System software, or systems software, is computer software designed to provide a **platform** to other software. Examples of system software include **operating systems**, computational science software, game engines, industrial automation, and software as service applications. In contrast to system software, software that allows users to do things like creating text documents, laying games, listening to music, or surfing the web is called **application software**.

system software 系统软件
software['sɒftweər] n. 软件
platform [ˈplætfɔ:m] n. 平台
operating systems 操作系统
application software 应用软件

Operating Systems or System Control Programs

An operating system (OS) is system software that manages computer hardware and software resources and provides common services for computer programs. It provides a platform (hardware abstraction layer) to run high-level system software and application software. A **kernel** is the core part of the operating system that defines an **API** for application programs (including some system software) and an interface to device drivers.

kernel [ˈkɜ:nl] n. 核心；要点
API=Application Program Interface 应用程序接口

Examples of Operating System

Microsoft Windows

Microsoft Windows is a family of **proprietary** operating systems designed by Microsoft Corporation and primarily targeted to Intel **architecture**-based computers, with an estimated 88.9 percent total usage share on Web-connected computers. The latest version is Windows 10.

proprietary [prəˈpraɪətri] adj. 专有的，专利的，n. 所有权
architecture [ˈɑ:kɪˌtektʃə] n. 结构，体系结构

In 2011, Windows 7 overtook Windows XP as the most common **version** in use. Microsoft Windows was first **released** in 1985, as an

version [ˈvɜ:ʃn] n. 译文，版本
release [rɪˈli:s] vt. 发布

operating environment running on top of MS-DOS, which was the standard operating system shipped on most Intel architecture personal computers at the time. In 1995, Windows 95 was released which only used MS-DOS as a **bootstrap**.

bootstrap ['bu:tˌstræp] n.引导程序

For backwards compatibility, Win9x could run real-mode MS-DOS and 16-bit Windows 3.x drivers. Windows ME, released in 2000, was the last version in the Win9x family. Later versions have all been based on the Windows NT kernel. Current client versions of Windows run on IA-32, x86-64 and 32-bit ARM microprocessors.

Server editions of Windows are widely used. In recent years, Microsoft has expended significant capital in an effort to promote the use of Windows as a server operating system.

Unix and Unix-like Operating Systems

Unix was originally written in assembly language. Ken Thompson wrote B, mainly based on BCPL, based on its experience in the MULTICS project. B was replaced by C, and Unix, rewritten in C, developed into a large, complex family of inter-related operating systems which have been **influential** in every modern operating system.

influential [ˌɪnflu'enʃl] adj. 有影响的

The Unix-like family is a **diverse** group of operating systems, with several major sub-categories including System V, BSD, and Linux. The name "UNIX" is a trademark of The Open Group which **licenses** it for use with any operating system that has been shown to conform to their definitions. "UNIX-like" is commonly used to refer to the large set of operating systems which resemble the original UNIX.

diverse [daɪ'vɜ:s] adj. 不同的

license ['laɪsəns] v. 批准，许可

Linux

The Linux kernel originated in 1991, as a project of Linus Torvalds, a university student in Finland. He posted information about his project on a newsgroup for computer students and programmers, and received support and assistance from **volunteers** who succeeded in creating a complete and functional kernel.

volunteer [ˌvɒlən'tiə] n. 义务工作者

Linux is Unix-like, but was developed without any Unix code, unlike BSD and its **variants**. Because of its open license model, the Linux kernel code is available for study and modification, which resulted in its use on a wide range of computing machinery from supercomputers to smart-watches. Although estimates suggest that Linux is used on only 1.82% of all "desktop" (or laptop) PCs, it has been widely adopted for use in servers and embedded systems such as cellphones. Linux has superseded Unix on many platforms and is used on most supercomputers including the top 385.

variant ['veəri:ənt] n. 变体

Mac OS

Mac OS (formerly "Mac OS X" and later "OS X") is a line of open core graphical operating systems developed, marketed, and sold by Apple Inc., the latest of which is pre-loaded on all currently shipping Macintosh computers. Mac OS is the successor to the original classic Mac OS, which had been Apple's primary operating system since 1984. Unlike its, Mac OS is a UNIX operating system built on technology that had been developed at NeXT through the second half of the 1980s and up until Apple purchased the company in early 1997. The operating system was first released in 1999 as Mac OS X Server 1.0, followed in March 2001 by a client version (Mac OS X v10.0 "Cheetah"). Since then, six more distinct "client" and "server" editions of Mac OS have been released, until the two were merged in OS X 10.7 "Lion".

.End.

Key Words

software n.软件　　platform n.平台
operating systems 操作系统　　system software 系统软件
application software 应用软件　　kernel n.核心；要点
API=Application Program Interface 应用程序接口
proprietary adj.专有的，专利的， n.所有权
architecture n.结构，体系结构
version n.译文,版本　　release vt.发布
bootstrap n.引导程序　　influential adj.有影响的
diverse adj.不同的　　license v.批准，许可
volunteer n.义务工作者　　variant n.变体

参考译文　技能1 系统软件——操作系统

系统软件，是为其他软件提供平台的计算机软件。系统软件包括操作系统、科学计算软件、游戏机、工业自动化和软件，是一种服务的应用。与系统软件相反，允许用户做诸如创建文本文档、放置游戏、听音乐或用Web浏览器来浏览Web的软件称为应用软件。

操作系统或系统控制程序

操作系统（OS）是管理计算机硬件和软件资源并为计算机程序提供公共服务的系统软件。

它为运行高级系统软件和应用软件提供了一个平台。内核是操作系统的核心部分，它定义了应用程序（包括一些系统软件）的接口和设备驱动程序的接口。

操作系统的示例

微软视窗软件

Microsoft Windows是一个专有操作系统的家族，它由微软公司设计，主要针对以英特尔公司体系结构为基础的计算机，估计使用量占共享网络连接计算机总量的88.9%。最新版本是Windows 10。

2011年，Windows 7取代了Windows XP，成为最常见使用的版本。微软视窗系统于1985年首次发布，作为一项操作环境运行在MS-DOS之上，这是当时大多数英特尔架构个人电脑上的标准操作系统。1995年，Windows 95被发布，只使用MS-DOS作为引导程序。

为了向后兼容，Win9x可以运行实模MS-DOS和16位Windows 3.x驱动程序。2000年发布的Windows ME是在Win9x家族中的最后一个版本。后来的版本都基于Windows NT内核。当前Windows客户端版本在IA-32，x86-64和32位ARM微处理器上运行。

Windows的服务器版本被广泛使用。近年来，微软花费了大量资金来推广Windows作为服务器操作系统的使用。

UNIX和类UNIX操作系统

UNIX最初是用汇编语言编写的。Ken ThompsonB主要基于BCPL，基于它在MULTICS的经验项目。B被C取代，UNIX被重写成C，后来UNIX发展成一个大型、复杂的相互关联的每个现代操作系统中都有影响力的操作系统家族。

类UNIX的家庭是一组不同的操作系统。几个主要的子类别有System V、BSD和Linux。“UNIX”的名称是开放组的一个标志，它为那些使用显示符合它们定义的操作系统颁发许可证。“类UNIX”通常用于指大型集合与原来的UNIX类似的操作系统。

Linux

Linux内核起源于1991年，是Linus Torvalds在芬兰还是大学生时候的一个项目，他面向计算机学生和程序员发布了有关这个项目的信息，并得到了志愿者的支持和帮助，并创建一个完整的功能内核。

Linux是类UNIX的，但它的开发没有任何的UNIX代码，不像BSD和它的变体。因为它的开放许可证模型，Linux内核代码用于学习和修改，这导致了它的使用从超级计算机到各种各样的计算机器smart-watches。尽管Linux估计在台式机(或笔记本电脑)的使用仅占1.82%，但它已被广泛用于服务器和嵌入式系统，如手机。Linux已经取代了许多平台上的UNIX，并且在大多数平台上使用超级计算机，包括前385。

Mac OS

Mac OS(以前的“Mac OS X”和后来的“OS X”)是由苹果公司(Apple Inc.)开发、推广和销售一条直线式开放核心图形操作系统，最新的一款产品已预装在目前所有出货的麦金塔电脑中。Mac OS是最初的经典Mac OS的继承者，自1984年以来一直是苹果的主要操作系统。与它的前身不同的是，Mac OS是一种UNIX操作系统基于NeXT技术在上世纪80年代后半段开发出来的，直到1997年初苹果公司收购了该公司。该操作系统于1999年首次发布为Mac OS X服务器1.0，紧接着在2001年3月发布了客户端版本(Mac OS X v10.0“Cheetah”)。从那以后，六个Mac OS的moredistinct“客户端”和“服务器”版本已经发布，直到这两个版本被合并到OS X 10.7“Lion”中。

Skill Two Application Software—OA

Application software helps you accomplish specific tasks. You can use application software to write letters, manage your **finance**, draw pictures, play games and so on. Application software is also called software, an application or a program. You can buy software at computer stores. There are also thousands of programs available on the Internet.

finance [fai'næns] n. 财务

Software you buy at a computer store is usually on a single **CD-ROM** disk, a **DVD-ROM** or several **floppy disks**. Before you use the software, you install, or copy the content of the disk or disks to your computer. Using a CD-ROM or DVD-ROM disk is a fast method of **installing** software.

CD-ROM 光盘只读存储器
DVD-ROM 数字化视频光盘存储器
floppy disk 软盘
installing n. 安装

When a manufacturer adds new **features** to existing software, the updated software is given a new name or a new **version** number. This helps people distinguish new versions of the software from older versions. Manufacturers may also create minor software updates, called patches, or improvements to software. A patch is also often referred to as a service pack.

feature [fi:tʃə] n. 属性
version ['və:ʃən] n. 版本

Bundled software is the software that comes with a new computer system or device, such as a printer. Companies often provide bundled software to let you start using the new equipment right away. For example, new computer systems usually come with word processing **spreadsheet** and **graphics** programs. Most softwares come with a built-in help feature and printed **documentation** to help you learn to use the software. You can also buy computer books that contain detailed, step-by-step instructions or visit the manufacturer's **web site** for more information about the software. The **OA** is technology that reduces the amount of human effort necessary to perform tasks in the office.

bundle ['bʌndl] v.绑定
spreadsheet['spredʃi:t] n.电子表格
graphic ['græfik] n.图形
documentation [ˌdɔkjumen'teiʃən] n. 文件
web site 网站
OA office automation 办公自动化化

Today's businesses have a wide variety of the OA technology at their **disposal**. Such as, data processing, **word processing**, **graphic**, image, voice and networking. The widespread use of the OA technology began in the workplace in offices, banks and factories etc. The development of the OA system has been **synchronizing** with the development of data, information and knowledge. The first OA system is mainly used to deal with data; while the second OA system is chiefly used to deal with information; but the third generation is used to deal with knowledge. These three stages in the development have accomplished the **leap** from the data processing to the information processing, as well as the leap from the information processing to the knowledge processing. In the development of the OA system, its scope for using increasingly widened, the dealings with the content promoted step by step, the system function was perfect ultimately.

disposal [di'spəuzəl] n. 处理
word processing 文字处理
graphic ['græfik] n.图形
synchronize ['siŋkrənaiz] v. 同步
leap[li:p] n. 飞跃

The new version of the Office suite has been publicly available from Microsoft. The new version of office suite will have build-in native **XML** support for Excel and Access. Excel users will also find new features for working with web components so that the used Excel might **automatically** publish a **web page**, or a **pivot** table/chart. **Smart Tags** appear on the screen while you work, offering information on completing tasks faster. For example, Smart Tags display **AutoCorrect, AutoFormat** and **Paste** options that can keep users from searching through **menus**. There are also many features found in the new version of office designed to help when data losing has occurred. Excel, PowerPoint, and Outlook will join Word in offering AutoRecover feature, which automatically saves documents and data at **intervals**.

XML(eXtensible Markup Language) 可扩展标记语言
automatically [ɔ:tə'mætikli] adv. 自动地
web page 网页
pivot ['pivət] adj. 重要的
Smart Tag 智能标记
AutoCorrect 自动纠错
AutoFormat 自动格式化
paste [peist] vt. 粘贴
menu ['menju:] n. 菜单
interval ['intəvəl] n. 间隔

Today's organizations have a variety of the OA hardware and software components at their disposal. The list includes telephone computer systems, electronic mail, word processing, **desktop publishing**, database management system, two-way cable TV, office-to-office satellite broadcasting, on-line database service, and voice recognition and synthesis. Each of these components is intended to automate a task or function that is presently performed manually.

desktop publishing 桌面印刷系统

.End.

tag n. 标记
graphic n. 图形
version n. 版本
documentation n. 文件
application software 应用软件
web page 网页
floppy disk 软盘
DVD-ROM 数字化视频光盘存储器
AutoFormat 自动格式化
data processing 数据处理
XML(eXtensible Markup Language) 可扩展标记语言
menu n. 菜单
installing n. 安装
spreadsheet n. 电子表格
paste vt. 粘贴
web site 网站
office automation 办公自动化
CD-ROM 光盘只读存储器
AutoCorrect 自动纠错
system software 系统软件
word processing 文字处理

参考译文 技能2 应用软件——办公自动化

应用软件可以帮助你完成特定的任务，你可以用应用软件写信、管理财务、画画、玩游戏等等。应用软件也叫软件、应用程序或程序。你可以从计算机商店中买到软件，还可以从互联网上得到成千上万个软件。

你从计算机商店中购买的软件通常是简单的CD盘、DVD盘，或者是几张软盘。在你能够使用这些软件之前，需要安装或者将这些盘上的内容拷贝到计算机上。使用CD盘或者DVD盘是一种较快的安装软件的方式。

当制造商向已经存在的软件上面添加新的属性时，通常会给它起一个新的名字或者是新的版本号，这有助于人们把新版本的软件与旧版本软件区分开来。制造商也可以做一些小规模的软件升级，叫作补丁，也叫升级软件。补丁通常叫作服务包。

绑定软件是指那些与一些新的计算机系统或者设备（例如打印机）一起售出的软件。一些公司通常会提供绑定软件，以便让你马上开始使用新设备。例如，新的计算机系统通常会附带一些文字处理软件、电子表格软件以及图形软件。大多数软件带有一个内置的帮助功能和打印文档来帮助你学习使用软件。你也可以买包含详细信息的电脑书，一步一步地来学习使用软件，也可以访问制造商的网站以获取更多的信息。办公自动化技术可以减少人们在办公室里执行任务时所必须完成的工作量。

今天的企业有各种技术的办公软件，如数据处理、文字处理、图形、图像、声音和网络。办公自动化技术的广泛应用始于办公室办公、银行和工厂等。办公软件系统的发展和数据、信息和知识的发展同步。第一代办公软件系统主要用于处理数据；第二代办公软件系统主要用来处理信息；而第三代则用于处理知识。这三个阶段的发展已经完成了从数据处理到信息处理，以及从信息处理到知识处理的飞跃。随着办公自动化系统的发展，其使用范围日益扩大，处理内容逐步提升，系统功能逐步完善。

微软已经推出新版Office，新版Office为Excel和Access内建了XML支持。Excel用户在使用web组件时也会发现新功能，可以自动地发布网页或重要图标。用户工作时智能标记会出现在屏幕上，提供信息并使任务完成得更快。例如，智能标记显示自动纠错、自动格式化以及粘贴行，从而使用户不必去搜寻菜单。当产生数据丢失后，新版本Office软件中会有许多新功能处理相关问题。Excel、PowerPoint和Outlook将同Word一样提供自动恢复功能，每间隔一段时间便会自动存储文档和数据。

如今的机构已经配置了各种各样的办公自动化硬件和软件，包括电话及计算机系统、电子邮件、文字处理、桌面印刷系统、数据库管理系统、双向电缆电视、办公室对办公室的卫星广播、联机数据库服务、声音识别及合成系统。这些配备都力图使目前需手工完成的任务或功能自动化。

Fast Reading One　Computer Development

Wearable computers are the next wave of portable computing and they will go way beyond laptops.

What's a Wearable?

Imagine watching a movie projected through your eyeglasses onto a virtual screen that seems to float in your field of vision. Or imagine working on an automobile, an airplane, or an underwater mission and reading an instruction manual, communicating with co-workers, or inputting data via computer—all without lifting a finger from the task at hand.

To date, personal computers have not lived up to their name. Most machines sit on the desk and interact with their owners for only a small fraction of the day. Smaller and faster notebook computers have made mobility less of an issue, but the same staid user paradigm persists.

Pic 2.2 Wearable computer

Wearable computer (Pic 2.2) hopes to shatter this myth of how a computer should be used. A wearable computer is a very personal computer. A person's computer should be worn, such as eyeglasses or clothing are worn, and interact with the user based on the context of the situation. With heads-up displays, unobtrusive input devices, personal wireless local area networks, and a host of other context sensing and communication tools, the wearable computer can act as an intelligent assistant.

Micro Optical Corp., based in Westwood, Mass., has designed two models of an eyeglass display—one that clips onto the side of the user's glasses, and one integrated directly into the eyewear. Though the user still needs a CPU — a laptop, a wearable computer, or even a DVD player or cell phone — the monitor or screen is actually projected through the user's eyeglasses.

Gerg Jenkins, sales manager at Micro Optical, says the LCD is positioned near the user's temple. A projected image passes through the lenses of regular eyeglasses, bounces off a mirror, and displays the illusion of a full-size monitor floating in front of the user's face. The display weighs less than an ounce, so it's much more comfortable than some of the earlier head-mounted displays.

WetPC

Called the WetPC① (Pic 2.3), it comprised a miniature personal computer with a mask-mounted virtual display and a novel one-handed controller—called a Kord□ Pad②. The computer was mounted in a waterproof housing on the diver's air tank. A cable from it was connected to the waterproof virtual display which presented the diver with a high contrast display "floating" in the field of view. A second cable was connected to the Kord Pad, a 5-key device which the diver could hold in either hand and which was used to control the computer by pressing single or multiple keys. A Graphical User Interface (GUI) shows the user which key (or keys) to press. The GUI facilitates the wear ability and usability of the WetPC—underwater computer. It was the result of several years of research into interface design and functionality. Rather like playing the piano, the user can interact with the computer in a very natural way—the diver can access and record information with one hand, even while swimming.

Pic 2.3 WetPC

The WetPC can help salvage-divers, maritime archeologists, and police divers find objects, record or look for information, or simply monitor their location at all times. Scientists could use the unit for mapping and monitoring coral reefs. Navy divers could use the WetPC to search for mines and other unexplored devices.

The WetPC has applications in education, recreation, and tourism as well. Divers can use the WetPC to navigate reefs and create a digital guided tour for underwater tourists.

The Challenges of Wearable Computer

The creation of wearable computer is not without challenge — primarily, how to produce an affordable product. Most units aren't priced for the average consumer. On the development side, designers and engineers continually strive to increase the functionality and comfort of wearable computers.

"One of the big challenges for us is to continue to make them user-friendly — to maintain usability as the functionality goes up and the size goes down," says Danny Cunagin, president of Logic.

And most wearable computers use standard desktop computer specifications — Pentium processors, Windows operating systems, creating a user interface that is easy, unobtrusive, and comfortable to use.

.End.

① WetPC:湿PC，一种可在水下使用的计算机。

② Kord Pad：五键考德机，一种计算机输入设备。

参考译文 计算机的发展

可佩带式计算机是便携式计算机的下一个浪潮，而且将远远超过膝上型计算机。

什么是可佩带式计算机？

想象一下，电影通过你的眼镜投影到虚拟屏幕上来看会是什么样子？好像电影画面就浮现在你的眼前！或者再想象一下，你正在开车、开飞机或执行海底任务，读到了操作指令，和对方进行交流，把数据输入到计算机——所有的这一切都没有动一下手指头。

时至今日，个人计算机已无法再胜任它的名字了。许多计算机被放置在桌子上，只在一天中很短的时间内被它们的主人使用。体积更小、速度更快的笔记本使便携不成问题，但是，固定的用户模式依然未变。

可佩带式计算机（图2.2）有望突破这种使用模式。可佩带式计算机是绝对的个人计算机。这种个人计算机将被随身佩带，就像眼镜和衣服一样。根据环境的不同，它可以与用户进行相应方式的互动。可佩带式计算机具有平视显示器、不引人注意的输入设备、个人无线局域网络、许多相关的感知和交流工具，可以充当一个智能助手。

位于美国马萨诸塞州韦斯特伍德的Micro Optical公司已经设计出两种基本的眼镜显示器模型：一种夹在用户的眼镜边上，另一种直接与眼镜结合在一起。尽管用户仍需要一个CPU，可以是笔记本、可佩带式计算机，甚至可以是DVD或移动电话，但它们的显示器或屏幕已经可以通过用户的眼镜进行投影了。

Micro Optical公司的销售经理Gerg Jenkins说液晶显示器应放置在用户的太阳穴附近。通过眼镜镜片投影的图像，从镜片上弹起，在用户的面前浮现出全屏的幻觉似的影像。这个显示器的重量不足一盎司，因此它比以前那种戴在头顶上的显示器舒服多了。

湿PC

所谓湿PC（图2.3），是一个小型的个人计算机，包含一个嵌在面罩内的显示器和一个独特的可由一只手控制的控制器（五键考德机）。这个计算机嵌在潜水员防水的氧气罐里。计算机的一条电缆连接到防水的显示器上，在潜水员眼前形成一个高对比度的“漂浮”画面；另一条电缆连接到五键考德机——一个只有五个键的设备上。潜水员可以用任一只手拿着，并且可以通过按一个或多个键来控制计算机。图形用户界面告诉用户该按哪一个键。图形用户界面的应用增强了湿PC的水下可佩带性和可用性，这是几年来对界面设计和功能性研究的结果。就像弹钢琴一样，用户可以用一种非常自然的方式操作计算机,即使正在游泳，潜水员也可以用一只手存取和记录信息。

湿PC能够帮助海上救助潜水员、海上考古学家和海上警察寻找目标，记录或查找信息，或实时监测目标的位置。科学家能够用它来勘测和监视珊瑚礁。海上潜水员能够利用湿PC来寻找矿藏，开采其他未被发现的宝藏。

湿PC还可以应用在教育、开发和旅游等方面。潜水员能够利用湿PC在海底礁石间领航，为水下的游客指路。

可佩戴式计算机所面临的挑战

可佩戴式计算机的产生是具有挑战性的，主要是如何生产出不太昂贵的计算机。许多器件对于普通用户来说太贵了。在开发方面，设计者和工程人员正在继续努力提高其功能性和舒适性。

“对于我们而言，一个巨大的挑战是在保证实用性、功能不断增强和体积不断减小的同时，继续使可佩戴式计算机的用户界面更友好。”Logic董事长Danny Cunagin说。

当前许多可佩戴式计算机都使用标准的台式机的规格——奔腾处理器、Windows操作系统，创建了一个简单、细致而且舒服的用户界面。

Fast Reading Two Linux

Many users get annoyed by everyday troubles that still affect Windows systems. Whether it is a failing program, so why not use Linux instead?

Linux is a freely available and distributable look-alike of UNIX developed primarily by Linus Torvalds(Pic 2.4) at the University of Helsinki in Finland. Linux was further developed with the help of many UNIX programmers and wizards across the Internet.

Pic 2.4 Linus Torvalds

Linux is small, fast, and flexible. Linux has been publicly available since about November 1991. Version 0.10 went out at that time, and version 0.11 followed in December 1991. There are very few bugs now, and in its current state, Linux is most useful for people who are willing to port and write new codes.

Linux is in a constant state of development. Linux is cheaper to get than most commercially available UNIX systems. If you have the patience and access to the Internet, the only price you pay for Linux is your time. For a nominal fee of anywhere from U. S. ＄30 to U. S. ＄200, you can save yourself some time and get CD-ROM or floppy-disk distributions from several commercial vendors. In my opinion, the most important advantage of using Linux is that you get to work with a real kernel. All the kernel source code is available for Linux, and you have the ability to modify it to suit your needs. Looking at the kernel code is an educational experience in itself.

There are different options in the Linux space. Red Hat and SuSe are two of the most popular ones. Red Hat Linux is a popular version of Linux that comes with the GNOME desktop environment. GNOME displays pictures on the screen to help you perform tasks. After that, selecting the software you want to use will also consume some time, but again you can be sure to find something that fits your needs.

Unlike Windows, most Linux distributions have a very modular structure, which means that you can choose to install a very slim basic system and a few, specialized modules only. As a result, a minimum installation does not require more than a few hundred megabytes.

Linux is a free UNIX clone that supports a wide range of software such as Xwindow systems, GNU, C/C++ Compiler and TCP/IP. It's a versatile, very UNIX—like implementation of UNIX, freely distributed by the terms of the GNU' General Public License. Linux is also very closely compliant with the OPOSIX.1 standard, so porting applications between Linux and UNIX systems is a snap.

Linux is a clone of the UNIX operating system that runs on Intel 80x86-based machines, where x is 3 or higher. Linux is also very portable and flexible because it has now been ported to DEC Alpha, PowerPC, and even Macintosh machines. And progress is being made daily by Linux enthusiasts all over the world to make this free operating system available to all the popular computing machines in use today. Because the source code for the entire Linux operating system is freely available, developers can spend time actually porting the code, regardless of the payment of licensing fees.

Documentation for the many parts of Linux is not very far away, either. The Linux Documentation Project (LDP) is an effort put together by many dedicated and very smart individuals to provide up-to-date, technically valuable information. All of this LDP information can be found on the Internet at various Linux source repositories. Each "HOWTO" document for Linux is the result of effort from many Linux enthusiasts. The original authors of these documents are usually also the core Linux developers who have put in hours of time and effort while struggling with new features of Linux. These individuals are the ones who deserve the credit.

.End.

参考译文 Linux

许多用户为Windows操作系统每天都在出问题而感到烦恼。也许这是一个有缺陷的程序，为什么我们不用Linux来代替它呢？

Linux是一个酷似UNIX而且可以免费获取的操作系统，最初是由芬兰赫尔辛基大学的Linus Torvalds（图2.4）研制的。后来Linux通过因特网，在许多UNIX程序设计人员和精英的帮助下得到了进一步开发。

Linux具有小、快、灵活的特点。大约在1991年11月，Linux已开始公开发行。那时推出了

0.10版，接着在当年的12月份推出了0.11版。当前版本的Linux错误极少，对于那些想移植和编写新代码的人来说，是非常有用的。

Linux处于不断开发之中。它比商业化的UNIX系统便宜。如果你有耐心访问因特网，只要付出一定的“时间”便可以购买到Linux。在任何地方，正常只需支付30美元至200美元，你便可以节约时间从一些商业售货机构获得CD-ROM，或是一些软盘的发行版。在我看来，使用Linux的最大好处就是可以用一个真正的内核工作。所有的内核源代码Linux都可使用，并能加以修改以满足用户的需要。研究内核代码本身就是自学的好方法。

Linux有不同的版本可供选择。Red Hat和SuSe是两种最流行的版本。Red Hat Linux是Linux的流行版本，它拥有GNOME桌面环境。GNOME在屏幕上显示图形来帮助你执行任务。此外，选择你希望用的软件将要耗费一些时间，但是你一定会找到满足你需求的产品。

与Windows不同，大部分的Linux发行版具有非常模块化的结构，这意味着你可以仅选择安装非常小的基本系统和几个专门模块。其结果是，一个最小的安装不会要求多于几百兆的空间。

Linux是免费的UNIX的克隆，它支持的软件十分广泛，如Xwindow系统、GNU、C/C++编译器和TCP/IP等。它是通用的，实现起来非常像UNIX，并根据GNU的公用许可证（GPL）协议免费发行。Linux非常接近POSIX.1标准，所以在Linux与UNIX间移植应用程序非常快捷。

Linux是UNIX操作系统的克隆，它运行在基于Intel 80x86处理器的机器上，这里的x大于或等于3。Linux非常易于移植，灵活性很强，因为它现在已经被移植到如DEC Alpha、PowerPC，甚至是Mac机上了。全世界的Linux爱好者每天都在取得进展，使这个免费的操作系统能运行在当今所有流行的机器上。因为全部的Linux操作系统的源代码都是免费的，所以开发者实际上只需花费时间来移植代码，不用考虑支付受益许可费的问题。

Linux许多部分的文档还远未流传开。Linux的文档计划（LDP）是把众多富有奉献精神和极具智慧的个人的努力集中起来，以提供最新的技术和最有价值的信息的一项计划。所有的LDP文档信息都能在因特网上的各种Linux储存库中找到。Linux的每个“HOWTO”文档都是Linux爱好者共同努力的结晶。这些文档的原作者，通常也是花费了许多时间和精力的核心Linux的开发者，为Linux的新特性而奋斗着。这些人理应享有赞誉。

Exercise

Ex 1 What is an operating system? Try to give a brief summary according to the passage.

Ex 2 Fill in the table below by matching the corresponding Chinese or English equivalents.

Kernel	
	应用程序接口
interface	
	操作系统
application software	
	平台

Ex 3 Choose the best answer to the following questions according to the text we learnt.

1. An OS consists of one or more ________ that manage the operations of a computer.
   A. interfaces
   B. programs
   C. kernel techniques
   D. account numbers
2. ________ is estimated 88.9 percent total usage share on Web connected computers.
   A. platform
   B. Media Player
   C. Graphical User Interface
   D. Microsoft Windows
3. The latest version of Microsoft is ?
   A. Windows XP
   B. Windows 7
   C. Windows 10
   D. DOS
4. System software provide a________to other software.
   A. platform
   B. Plug and Play
   C. Media Player
   D. Windows 95
5. The following is not of the Examples of operating system?
   A. Windows
   B. Unix
   C. Linux
   D. 3DMAX

Part B Practical Learning

Training Target

In this part, our target is to train the students how to use specialized English knowledge to finish professional tasks in English environment. It can achieve the purpose of using professional English through these tasks.

Task One Download the Software (Windows 7 and Application Software) from the Internet

The first task is to download necessary software from the Internet. In this task, students must know the necessary software needed to finish the whole task.

There is some information about the download software. The information can help students finish the task.

The needed software can be downloaded through web page : www. baidu.com or www. google. com (Pic 2.5, Pic 2.6).

Pic 2.5

Pic 2.6

You can click on the relevant options to enter into relevant pages for details (Pic 2.7,Pic 2.8).

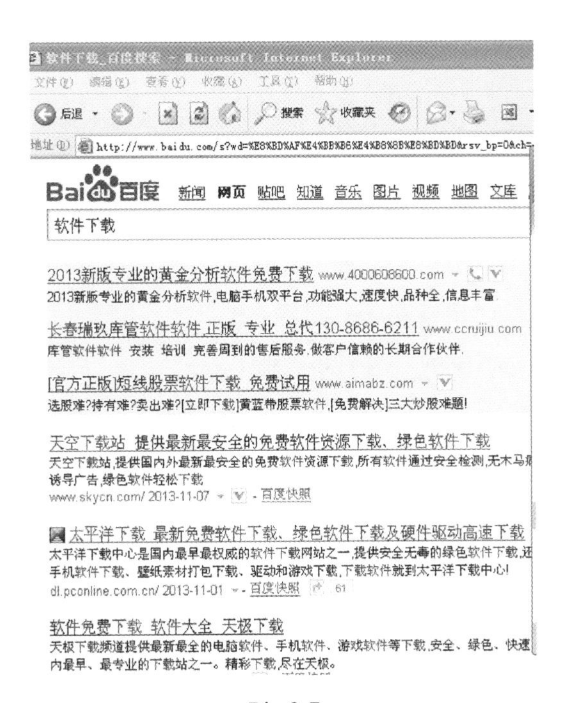

Pic 2.7

Pic 2.8

Task Two Install the Software

In this task, students must install the needed software which they download from the Internet. Just run the setup program and follow the prompts (Pic 2.9).

Pic 2.9

Part C Occupation English

Training Target

In this part, students are supposed to practice the dialogue, and act as staff of service center. Offering help to the customers in installation process is one of the common tasks in their future job.

Installing an Operating System 安装操作系统

Post: Computer Customer Service Staff（岗位：计算机客服人员）

安装操作系统是每个人都会遇到的问题，对计算机专业的学生来说，为顾客安装操作系统是日常最重要的工作之一。

A : Hello, how can I install Windows 7 on my computer? 请问，怎样在计算机上安装Windows 7操作系统？

B : I'm very glad to help you with the installation. To install an operating system, you should first clean up the primary partition, second boot your computer with the Windows 7 Installation Disk, third format the primary partition C with the installer. And fourth, you just follow the instructions of the installer and do necessary changes until it is finished. I advise you to consult your User Manual in the process, because some steps can be very dangerous.

非常高兴向你提供安装帮助。如果要安装操作系统，你首先要把主分区清理干净，然后插入Windows 7安装盘启动你的电脑，然后用安装程序格式化C分区。之后，你只要跟随安装程序的指示，再做一些必要的修改就可以成功安装了。我建议你在安装过程中，参考一下用户手册，因为有些步骤会带来危险。

A : I am wondering whether you can tell me how to install Linux on my computer. 请问，怎样在计算机上安装Linux操作系统？

B : OK, you are welcome. First, prepare a clean partition for the operating system. Then, boot your computer with the first Linux Installation Disk, and then format that partition with a Linux supported system. After that, you can just follow the installation instructions. You don't need to change anything. 没问题。首先准备一张空白的磁盘分区。然后插入Linux的第一张安装盘启动你的电脑，接下来用一种适合Linux的文件系统来格式化该分区。然后只需要按照安装程序的指示做就可以了。无需做任何更改。

A : How can I set up a firewall for my Windows 7 system? 怎样为我的Windows 7操作系统设置防火墙啊？

B : Windows XP itself provides the internal Internet firewall. Enable it, and it will be OK. The process is rather easy. Go to the Control Panel, enter "Network Connections". Now right-click on your current connection, and select Properties. In the advanced tab, enable the Internet Connection Firewall. That's clear, isn't it? Windows 7 操作系统带有内置因特网防火墙，启用它就可以了。具体方法是，进入控制面板，打开网络连接，右击当前连接，选择属性，在高级选项中，启用互联网连接防火墙，清楚了吧？

A : Thank you. 谢谢。

Exercise: Practice the dialogue in pairs.

前缀/后缀由一个或几个字母组成，放在词根或单词之前/之后，组成一个新词。

(1) kilo-(前缀)：千

gram：克——kilogram：千克

meter：米——kilometer：千米

(2) sub-（前缀）：在……底下，子，次

head：标题——subhead：小标题，副标题

directory：目录——subdirectory：子目录

(3) en-（前缀)：使……

able：有才能的——enable：使能够

close：关闭——enclose：装入，附上

(4) -ment（后缀）：行为，状态，过程，手段及其结果

move：移动——movement：运动

judge：判断——judgement：判断

(5) -th（后缀）：动作，性质，过程，状态

true：真实的，真正的——truth：事实，真理

weal：福利，幸福——wealth：财富

Ex Translate the following words and try your best to guess the meaning of the word on the right according to the clues given on the left.

byte	字节（名词）	kilobyte	________
bit	位，比特（名词）	kilobit	________
way	路（名词）	subway	________
punish	惩罚（动词）	punishment	________
argue	争论，辩论（动词）	argument	________

large	大的（形容词）	enlarge	________
danger	危险（名词）	endanger	________
develop	发展（动词）	development	________
equip	装备（动词）	equipment	________
grow	生长（动词）	growth	________

Exercise

Ex 1 **What is software like? Try to give a brief summary of "*Application Software—OA*" in no more than five sentences.**

Ex 2 **What is application software? Try to give several examples.**

Ex 3 **What information does Smart Tags offer on the screen while you work?**

Ex 4 **Fill in the table below by matching the corresponding Chinese or English equivalents.**

version	
	应用软件
wearable computer	
	文件
paste	
	网页
data processing	
	光盘只读存储器
XML	
	办公自动化

Ex 5 **Choose the best answer to the following questions according to the text we learnt.**

1. Software is the set of ________ that tell a computer what to do.

   A. tools

   B. rules

   C. symbols

   D. instructions

2. Software can be categorized into two types: ________ and ________.

   A. bundled software, office software

   B. package software, custom software

   C. system software, application software

   D. freeware software, public-domain software

3. When you run your computer, ________ will be started first.

   A. operation system

   B. system software

   C. office software

   D. application software

4. When the user uses a new version of the Office to work, ________ offer(s) information and display(s) AutoCorrect, AutoFormat and Paste options.

   A. word processing system

   B. desktop publishing system

   C. Smart Tags

   D. voice recognition and synthesis

5. These three stages in the development of the OA system have accomplished the leap ________.

   A. from data processing to information processing

   B. from information processing to knowledge processing

   C. from data processing to knowledge processing

   D. A and B

Project Three

LAN Setup and Connecting it to the Internet

Part A Theoretical Learning

Part B Practical Learning

Part C Occupation English

Part A Theoretical Learning

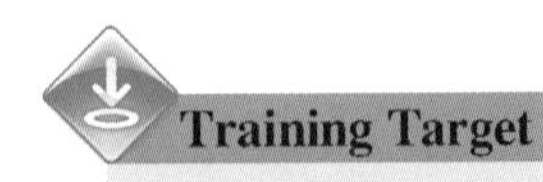

In this part, our target is to improve the speed of reading professional articles and the comprehension ability of the reader. We have marked specialized vocabulary key words in some paragraphs so that the reader can quickly grasp the main idea of the sentences and paragraphs.

Skill One Foundation of Network

Computer networks are data **communication** systems made up of hardware and software. **A network is a collection of computers and devices connected together via communication devices such as cable telephone lines, modems, or other means. Sometimes a network is wireless; that is, it uses no physical lines or wires. When your computer is connected to computer network, users can share resources, such as hardware devices, software programs, data, and information.** Sharing resources saves time and money. For example, instead of purchasing one printer for every computer via a network, the network enables all of the computers to access the same printer.

communication [kəˌmju:ni'keiʃən] n. 通信
via ['vaiə] prep. 通过，经过
wireless ['waiəlis] adj. 无线的
physical ['fizikəl] adj. 物理的

1. Types of Networks

Networks are often **classified** according to their **geographical** extent: LAN, MAN, WAN.

◇ **Local** Area Network (LAN)

A typical LAN spans a small area like a single building or a small campus and operates between 10 Mbps and 2 Gbps. Because LAN technologies cover short distances, they provide the highest speed connections among computers. **Ethernet** and **FDDI** are examples of standard LANs.

◇ **Metropolitan** Area Network (MAN)

MAN is a bigger version of a LAN in a city, it is smaller than a WAN but larger than a LAN. MAN is a public high-speed network, and runs at a speed of 100 Mbps or even faster, capable of voice and data transmission over a distance of up to 80 kilometers (50 miles).

◇ Wide Area Network (WAN)

WAN is sometimes called long haul network, providing communication over long distance. It can span more than one

classify ['klæsifai] v. 分类
geographical [dʒiə'græfikəl] adj. 地理的
local ['ləukl] adj. 地方的，当地的
Ethernet [iθə:net] n. 以太网
FDDI 光纤分布式数据接口
metropolitan [metrə'pɔlitən] adj. 首都的，大城市的

geographical area, often a country or continent. Usually WANs operate at slower speeds than LANs, and have much greater delay between connections. Typical speeds for a WAN range from 56Kbps to 155 Mbps. The Internet can be correctly regarded as the largest WAN in existence.

2. Topology

A network architecture can be described in two ways: client-server and **peer-to-peer**. A client-server network is a network comprised of several workstations and one or more servers. In client-server networks, an administrator can control the **privileges** of each user. A peer-to-peer network is a type of network where all computers on the network have the **potential** to share resources that they have control over. All computers on the network can potentially act as both a client and a server. Because of this fact, there is no central control of the network and therefore this type of network structure is considered to be less secure and harder to manage than the client-server architecture.

Topology defines the structure of the network. There are five major topologies in use today: Bus, Ring, Star, Tree, and **Mesh**. Each is used for specific network types, although some network types can use more than one topology.

topology [tə'pɔlədʒi] n. 拓扑结构
peer-to-peer 对等网络
privilege ['privilidʒ] n. 特权
potential [pə'tenʃl] n. 潜能，潜力
mesh [meʃ] n. 网孔，网丝

◇ Bus: The simplest topology is the Bus (Pic 3.1). In the Bus, all the devices on the network are connected to a common cable. Normally, this cable is terminated at either end, and can never be allowed to form a closed loop.

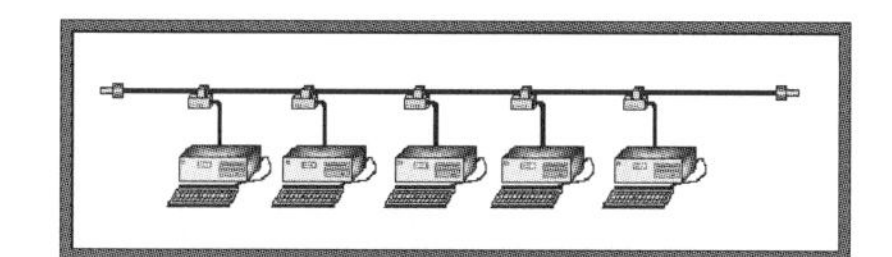

Pic 3.1 Bus topology

◇ Ring: A Ring topology (Pic 3.2) is very similar to a Bus. In a Ring, all the devices on the network are connected to a common cable which loops from machine to machine. After the last machine on the network, the cable then returns to the first device to form a closed loop.

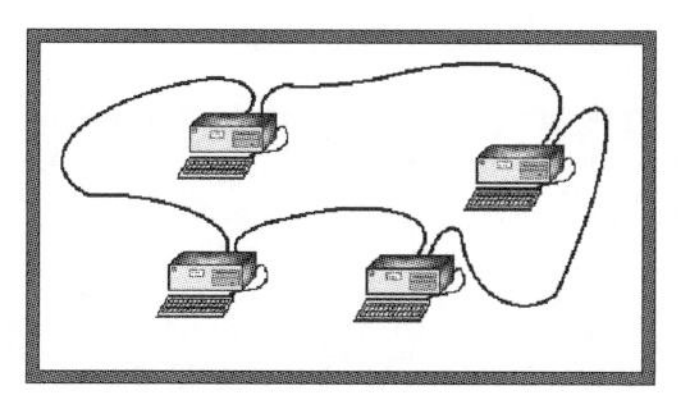

Pic 3.2 Ring topology

◇ Star: A Star topology (Pic 3.3) is completely different from either a Bus or a Ring. In a Star, each device has its own cable that connects the device to a common hub or concentrator. Only one device is permitted to use each port on the hub.

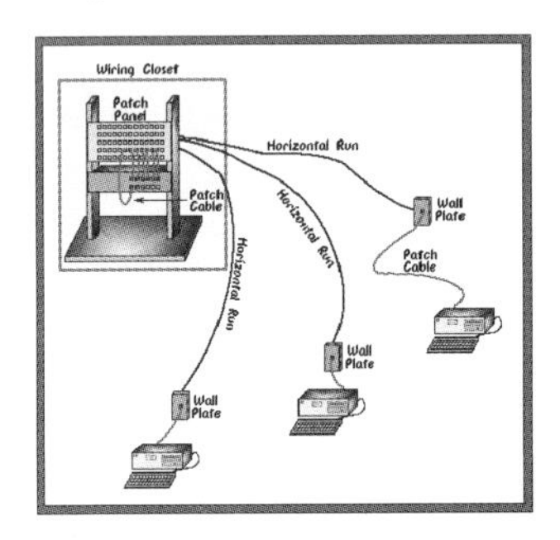

Pic 3.3 Star topology

◇ Tree: A Tree topology (Pic 3.4) can be thought of as being a "Star of Stars" network. In a Tree network, each device is connected to its own port on a concentrator in the same manner as in a Star. However, concentrators are connected together in a **hierarchical** manner.

hierarchical [haiə'rɑ:kikl]
adj. 分等级的

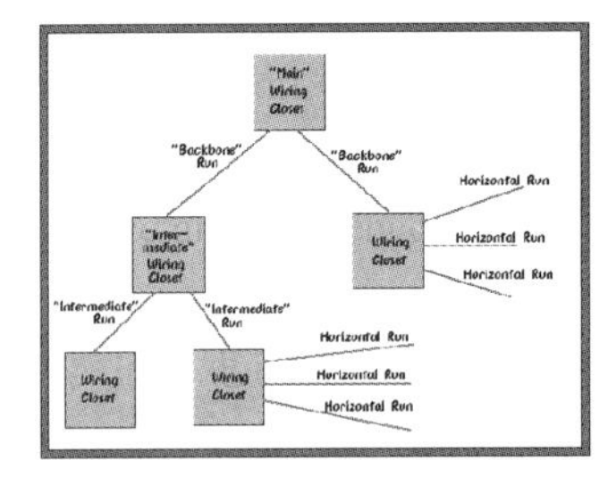

Pic 3.4 Tree topology

◇ Mesh: A Mesh topology consists of a network where every device on the network is physically connected to every other device on the network. This provides a great deal of performance and reliability, however the complexity and difficulty of creating one increases **geometrically** as the number of nodes on the network increases. Pic 3.5 shows an example of a four-node Mesh network.

geometrical [dʒiə'metrikl]
adj. 几何的，几何学的

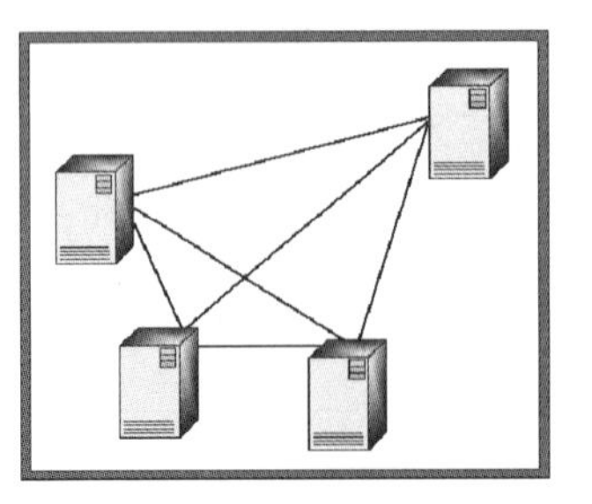

Pic 3.5 A four-node Mesh network

.End.

network n. 网络
LAN 局域网
WAN 广域网
FDDI 光纤分布式数据接口
peer-to-peer network 对等网络
Bus network 总线型网络
Star network 星型网络
Mesh network 网型网络
wireless adj. 无线的
MAN 城域网
Ethernet n. 以太网
client-server (C/S) 客户/服务器
topology n. 拓扑结构
Ring network 环型网络
Tree network 树型网络

参考译文 技能1 网络基础

计算机网络是由硬件和软件构成的数据通信系统。它通过通信设备比如电话线、调制解调器或其他方式将一系列的计算机和设备连接起来。有时，网络也可以是无线的，也就是不需要物理线路或电缆。当你的计算机连接到网络上，计算机用户就可以共享资源，比如硬件设备、软件程序、数据和信息。共享资源可以节省时间和金钱。举个例子吧，网络可以使处于同一网络内的计算机使用同一台打印机，而不用为网络中的每一台计算机都购买一台打印机。

1.网络类型

根据网络覆盖的地理范围，网络可以分为局域网、城域网和广域网。

◇ 局域网（LAN）

一个典型的局域网跨越的范围较小，诸如一幢大楼或者是一个小型校园，运行的速度在10Mbps到2Gbps之间。局域网技术覆盖较短的距离，从而提供了计算机间的最高速连接。以太网和FDDI就是典型的局域网。

◇ 城域网（MAN）

城域网是一种用于城市间的大型局域网，它比广域网(WAN)规模小，但比局域网(LAN)规模大。城域网是一种公用的高速的计算机网络，运行速度为100Mbps甚至更快，可以传输语音和数据，最大传输距离能达到80公里(50英里)。

◇ 广域网（WAN）

广域网有时称为远程网，它能提供长距离的通信。WAN的跨度不止一个地区，可以包含一个国家或一个州。广域网的运行速度比局域网低，而且在连接之间有较大的传输延时。广域网的典型速度在56Kbps到155Mbps之间。因特网是现存最大的广域网。

2.拓扑结构

网络体系结构有两种类型：客户/服务器模式和对等网络模式。客户/服务器网络是一个包括几个工作站和一台或多台服务器的网络。在客户/服务器网中，管理员能够控制每个用户的权限。在对等网络中，所有计算机都能共享它们控制的资源。网络中的计算机既是客户又是服务器。正是因为没有网络的集中控制，与客户/服务器模式相比，这种网络结构被认为缺少安全性，并且更难管理。

拓扑定义了网络的结构。目前主要有5种拓扑结构：总线型、环型、星型、树型和网型。一种拓扑结构适用于一个具体的网络，当然一些网络可使用不止一种的拓扑结构。

◇ 总线型：最简单的拓扑结构就是总线型（图3.1）。在总线型网络中，所有的网络设备都连接到一条公用电缆上，这条电缆在两端终结，并且永远不能形成一个封闭环路。

◇ 环型：环型拓扑（图3.2）非常类似于总线型。在环型网络中，所有的网络设备都连接到一条公共缆线上，缆线从一个设备连到另一个设备，再从最后一个设备连回到第一个设备，形成一个封闭的环路。

◇ 星型：星型拓扑（图3.3）完全不同于总线型或者环型。在星型网络里，每个设备都通过专用电缆连到集线器或中心控制器。一台设备只允许连到集线器的一个端口。

◇ 树型：树型拓扑（图3.4）可以被认为是一种“星型中的星型”的网络。在树型网络内，每个设备就像在星型网络中那样连接到次级集线器中的一个端口上，再由次级集线器连到上级集线器。

◇ 网型：在网型拓扑结构中，网络中的每个设备都与网络中其他所有设备有一条物理连接，在很大程度上提高了网络的健壮性和可靠性。但是随着网络节点的增加，复杂性和安装的难度也增大了。图3.5显示的是一个4个节点的网型网络。

Skill Two Network Devices

Network devices include all computers, media, interface cards and other equipments needed to perform data-processing and communications within the network. Let's look at some typical network devices.

◇ **Network Interface Card (NIC)**

The network interface card provides the physical connection between the network and the computer workstation. Most NICs are internal, with the card fitting into an **expansion** slot inside the computer. NIC is a major factor in determining the speed and performance of a network. It is a good idea to use the fastest network card available for the type of workstation you are using. NICs are considered Layer 2 devices because each individual NIC throughout the world carries a **unique** code, called a Media Access Control (MAC) address. This address is used to control data communication for the host on the network.

expansion [ik'spænʃən] n. 扩展

unique [ju'ni:k] adj. 唯一的

◇ **Repeater**

One of the disadvantages of the type of cable that we primarily use, CAT5 UTP, is cable length. The maximum length for UTP cable in a network is 100 meters (approximately 333 feet). If you need to extend beyond the network limit, you must add a device to your network. This device is called a repeater. The repeater electrically **amplifies** the signal it receives and rebroadcasts it. Repeaters are networking devices that exit at Layer 1, the physical layer, of the OSI reference model.

repeater [ri'pi:tə]
n. 中继器

amplify ['æmplifai]
v. 放大，增强

◇ **Hub**

Generally speaking, hub is used when referring to the device that serves as the center of a network, as shown in Pic 3.6. The purpose of a hub is to **regenerate** and retime network signals. You will notice the characteristics of a hub are similar to the repeater's, which is why a hub is also known as multi-port repeater. The difference is the number of cables that connects to the device. Where a repeater typically has only two ports, a hub generally has from four to twenty or more ports. Whereas a repeater receives signal on one port and repeats it on the other, a hub receives signal on one port and transmits it to all other ports.

regenerate [ri'dʒenəreit]
v. 再生

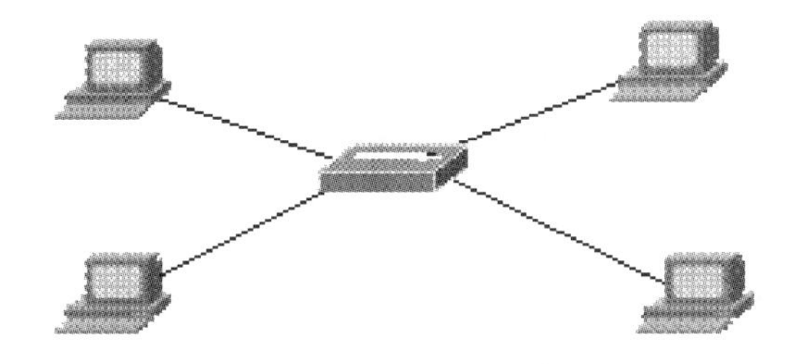

Pic 3.6 Hub as the center of a network

◇ **Bridge**

A bridge is a Layer 2 device that allows you to **segment** a large network into two smaller, more efficient networks. Bridges can be used to connect different types of cable, or physical topologies. They must, however, be used between networks with the same protocol. Bridges are **store-and-forward** devices. Every networking device has a unique MAC address on the NIC. Bridges **filter** networking traffic by only looking at the MAC address. Therefore, they can rapidly forward traffic representing any network layer protocol. Because bridges look only at MAC addresses, they are not concerned with network layer protocols. Consequently, bridges are concerned only with passing or not passing **frames**, based on their destination MAC addresses.

segment ['segmənt] v. 分割

store-and-forward 存储转发

filter ['filtə] v. 过滤

frame [freim] n. 帧

◇ **Switch**

A switch is a Layer 2 device just as a bridge is. In fact, a switch is sometimes called a multi-port bridge, just like a hub is called a multi-port repeater. Switches, at first glance, often look like hubs. Both hubs and switches have many connection ports. The difference between a hub and a switch is what happens inside the device. Switches make a LAN much more efficient. They do this by "switching" data only out the port to which the proper host is connected. In contrast, a hub sends the data out all its ports so that all the hosts have to see and process all the data. Pic 3.7 shows a switch and the symbol for a switch.

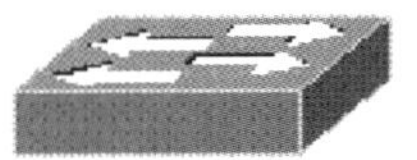

Pic 3.7 A switch and the symbol for a switch

◇ **Router**

The router is the first device you work with that is at the OSI network layer, otherwise known as Layer 3. The router makes decisions based on network addresses as opposed to individual layer 2 MAC addresses. The purpose of a router is to examine incoming packets, choose the best path for them through the network, and then switch them to the proper outgoing port. Routers are the most important traffic regulating devices on large networks. They enable virtually any type of computer to communicate with any other computer anywhere in the world! Pic 3.8 shows a router and the symbol for a router.

Pic 3.8 A router and the symbol for a router

◇ **Gateway**

A gateway can translate information between different network data formats or network architectures. It can translate TCP/IP to AppleTalk, so that computers supporting TCP/IP can communicate with Apple brand computers. Most gateways operate at the application layer.

◇ **Transmission Media**

Transmission media can be transferred in wired or wireless method. The basic wired media are **twisted pair**, **coaxial** cable, and **optical fiber**. Wireless media have **terrestrial** microwave, satellite microwave and broadcast radio. Networking media are considered Layer 1 components of OSI model.

.End.

twisted pair n. 双绞线
coaxial [kəu'æksiəl] adj. 同轴的
optical ['ɔptikəl] adj. 光学的
fiber ['faibə:] n. 光纤
terrestrial [tə'restriəl] adj. 陆地的

Key Words

repeater n. 中继器

hub n. 集线器

bridge n. 网桥

switch n. 交换机

router n. 路由器

cable n. 电缆

frame n. 帧

segment v. 分割

gateway n. 网关

transmission media n. 传输介质

twisted pair n. 双绞线

coaxial cable n. 同轴电缆

optical fiber n. 光纤

NIC (Network Interface Card) 网络接口卡（网卡）

MAC (Media Access Control) address 媒体访问控制地址

OSI (Open System Interconnect) reference model 开放式系统互联参考模型

STP (Shielded Twisted-Pair) 屏蔽双绞线

UTP (Unshielded Twisted-Pair) 非屏蔽双绞线

参考译文 技能2 网络设备

网络设备包括计算机、介质、网卡和网络中用于数据处理和通信的各种设备。下面就让我们看一些典型的网络设备吧！

◇ 网卡（NIC）

网卡为工作站和网络之间提供了一个物理连接，许多网卡都是内置的，它插在计算机内的扩展槽中。网卡是影响网络速度和性能的一个主要因素，应尽可能为你的工作站配备最快速的网卡。网卡被认为是第二层的设备，因为世界上的每一块网卡都携带有一个被称为媒体访问控制地址（即MAC地址）的编码。这个地址被用来控制网上主机的数据通信。

◇ 中继器

我们主要使用的电缆（CAT5 UTP）的缺点是电缆长度有限制。在一个网络中UTP电缆的最大长度是100米（大约333英尺）。如果你想要延伸网络并超过这个界限，就必须在网络中添加一种设备。这种设备被称为中继器。中继器将收到的信号进行放大后会再次传播。中继器是工作在OSI参考模型的第一层——物理层的网络设备。

◇ 集线器

一般来说，当我们需要一个设备作为网络中心时，就要利用集线器，如图3.6所示。集线器的作用是对网络信号进行再生和重新定时。你会注意到，集线器的特点与中继器极为相似，这就是集线器也被称为多端口中继器的原因。它们的区别是与设备相连接的电缆的数量不同。典型的中继器只有两个端口，而集线器的端口通常可有4个、20个甚至更多。中继器从一个端口接收信号，重发到另一个端口上；而集线器从一个端口接收到信号后，却将它传送到其他所有的端口上。

◇ 网桥

网桥是第二层的设备，它允许你把一个大的网络分成两个更小、更高效的网络。网桥可以用来连接不同传输介质或不同拓扑结构的网络。不过，它们必须在使用相同协议的网络之间使用。网桥是一种存储转发设备。每一个联网设备的网卡上都具有唯一的MAC地址。网桥仅根据MAC地址来过滤网络流量。因此，它们能够迅速地将代表着任何网络层协议的流量向前传送。由于网桥只关注MAC地址，它们就不注重网络层的协议。因此，网桥只根据帧的目的MAC地址关注帧是否能够通过。

◇ 交换机

交换机像网桥一样也是第二层的设备。实际上，交换机经常被称为多端口网桥，就像集线器被称为多端口中继器一样。粗看起来，交换机和集线器很像。无论是集线器还是交换机都有许多端口，它们的区别在设备的内部。交换机使LAN更有效。它们只将数据“交换”到相应主机的端口。相反地，集线器却将数据发送到它的全部端口上，以至于所有的主机不得不看到并且处理所有的数据。图3.7显示的是一台交换机和它的表示符号。

◇ 路由器

路由器是工作在OSI第三层——网络层的第一个器件。路由器依据网络地址做出决策，而不是依据第二层的MAC地址。路由器的作用是检查流入的数据包，为它们选择通过网络的最佳路径，然后把它们交换到适当的输出端口。在大型网络中，路由器是最重要的流量控制设备。它们几乎能够使计算机与世界其他任何地方的计算机进行通信。图3.8显示的是一台路由器和它的表示符号。

◇ 网关

网关能在不同的网络数据形式或者网络体系结构之间传输信息。网关能把TCP/IP协议转换成AppleTalk协议，因此支持TCP/IP协议的计算机能与苹果牌计算机进行通讯。大多数网关工作在应用层。

◇ 传输介质

传输介质可以是有线的也可以是无线的。基本的有线传输介质是双绞线、同轴电缆和光纤。无线介质包括陆地微波、卫星微波和无线电波。网络介质被认为是OSI模型第一层的器件。

Fast Reading One TCP/IP Protocols

A wide variety of network protocol models exist, which are defined by many standard organizations worldwide and technology vendors over years of technology evolution and development. One of the most popular network protocol model is TCP/IP, which is the heart of Internetworking communications.

The name TCP/IP refers to a suite of data communication protocols. The name is misleading because TCP and IP are only two of dozens of protocols that compose the suite. Its name comes from two of the more important protocols in the suite, the Transmission Control Protocol(TCP) and the Internet Protocol(IP). TCP/IP originated out of the investigative research into networking protocols that the Department of Defense (DoD) initiated in 1969. In the early 1980s, the TCP/IP protocols were developed. In 1983, they became standard protocols for ARPANET. Because of the history of the TCP/IP protocol suite, it is often referred to as the DoD protocol suite or the Internet protocol suite.

TCP/IP protocol suite includes more than 100 protocols, now let's see some typical protocols.

◇ **Hypertext Transfer Protocol (HTTP)**

HTTP is an application-level protocol for distributed, collaborative, hypermedia information systems. It has been in use by the World Wide Web global information initiative since 1990. It is a stateless protocol which can be used for many tasks beyond its use for hypertext, such as name servers and distributed object management systems, through extension of its request methods, error codes and headers. A feature of HTTP is the typing and negotiation of data representation, allowing systems to be built independently of the data being transferred.

◇ **File Transfer Protocol(FTP)**

FTP enables file sharing between hosts. FTP uses TCP to create a virtual connection for control information and then creates a separate TCP connection for data transfers. The control connection uses an image of the TELNET protocol to exchange commands and messages between hosts.

◇ **Transmission Control Protocol (TCP)**

TCP supports the network at the transport layer. TCP specifies the format of the data and acknowledgement that two computers exchange to achieve a reliable transfer, as well as the procedure the computers use to ensure that the data arrives correctly. It specifies how TCP software distinguishes among multiple destinations on a given machine, and how communicating machines recover from errors like lost or duplicated packets. The protocol also specifies how two computers initiate a TCP stream transfer and how they agree when it is complete. Because TCP assumes little about the underlying communication system, TCP can be used with a variety of packet delivery systems, including the IP datagram delivery service. In fact, the large variety of delivery systems TCP can use is one of its strengths.

◇ **User Datagram Protocol (UDP)**

UDP supports the network at the transport layer. UDP provides the primary mechanism that application programs use to send datagrams to other application programs. UDP provides protocol ports used to distinguish among multiple programs executing on a single machine. UDP uses the underlying IP to transport a message from one machine to another, and provides the same unreliable, connectionless datagram delivery semantics as IP. It does not use acknowledgement to make sure messages arrive, it does not order incoming messages, and it does not provide feedback to control the rate at which information flows between the machines.

◇ **Internet Protocol (IP)**

IP provides support at the network layer. IP provides three important definitions. First, the IP protocol defines the basic unit of data transfer used throughout a TCP/IP Internet. Thus, it specifies the exact format of all data as it passes across a TCP/IP Internet. Second, IP software performs the routing function, choosing a path over which data will be sent. Third, IP includes a set of rules that embody the idea of unreliable packet delivery, meaning there is no guarantee that the data will reach the intended host. The datagram may be damaged upon arrival, out of order, or not arrive at all. IP is such a foundational part of the design that a TCP/IP Internet is sometimes called an IP-base technology.

◇ **Internet Control Message Protocol (ICMP)**

ICMP used for network error reporting and generating messages that require attention. The errors reported by ICMP are generally related to datagram processing. ICMP only reports errors involving fragment 0 of any fragmented datagrams. The IP, UDP or TCP layer will usually take action based on ICMP messages. ICMP generally belongs to the IP layer of TCP/IP but relies on IP for support at the network layer. ICMP messages are encapsulated inside IP datagrams.

Evolution of TCP/IP technology is interwined with revolution of the global Internet. With millions of users at tens of thousands of sites around the world depending on the global Internet as part of their daily work environment, we have passed the early stage of development in which every user was also an expert, and entered a stage which few users understand the technology. Despite appearances, however, neither the Internet nor the TCP/IP protocol suite is static. Researchers solve new networking problems, and engineers improve the underlying mechanisms. In short, TCP/IP technology continues to evolve.

.End.

参考译文 TCP/IP协议栈

许多国际标准化组织和技术开发商经过多年的技术开发和发展，形成了多种网络协议模型。最流行的是TCP/IP协议，它是互联网通信的核心。

TCP/IP指的是一组数据通信协议，它有时被人们误解，因为TCP和IP只是协议栈中的两个协议。这个名称来源于协议栈中两个重要的协议：传输控制协议(TCP)和网际协议(IP)。TCP/IP来源于1969年国防部（DoD）对网络协议的开发研究。20世纪80年代初期，TCP/IP得到了发展。1983年，它成为ARPA网的标准协议。由于TCP/IP协议栈的发展历史，它也经常被称作DoD协议或Internet协议栈。

TCP/IP协议栈包含了100多个协议，我们看看其中的几个典型协议：

◇ **超文本传输协议（HTTP）**

HTTP是一个应用层的协议，用来发布、协调超文本信息。从1990年起它一直为WWW万维网所使用。HTTP是一种通过请求方法、错误代码和报头的扩展格式来实现的无状态协议。它除了用于超文本传输之外，还用于很多其他的任务，例如，名称服务器和分布式的对象管理系统。HTTP的特点是允许系统独立地创建数据表示方法的类型和协商能力，而不依赖于所要传输的数据。

◇ 文件传输协议（FTP）

FTP协议使在主机之间共享文件成为可能。FTP使用TCP为控制信息建立虚拟连接，然后为数据传输创建一个单独的TCP连接。控制连接使用TELNET协议在主机之间交换命令和消息。

◇ 传输控制协议（TCP）

传输控制协议(TCP)工作在传输层。TCP指定了两台计算机之间为了进行可靠传输而交换的数据和确认信息的格式，还指定了计算机为了确保数据的正确到达而采取的步骤。该协议规定了TCP软件如何识别给定机器上的多个目标，及如何对类似分组丢失和分组重复这样的错误进行恢复。该协议还指出了如何在计算机之间实现发起TCP数据流的传输，以及完成后计算机如何同意开始传输数据。由于TCP对底层通信系统没有什么特殊要求，因此可用于包括IP数据报交付服务在内的多种数据包传输系统，这正是其强大功能的体现。

◇ 用户数据报协议（UDP）

UDP工作在传输层。UDP提供应用程序之间传输数据报的基本机制。UDP提供的协议端口能够区分在一台机器上运行的多个程序。UDP使用底层的IP协议在各个机器之间传输报文，提供和IP一样不可靠、无连接的数据报传输服务。它没有使用确认机制来确保报文的准确到达，没有对传入的报文排序，也不提供反馈信息来控制机器之间信息传输的速度。

◇ 网际协议（IP）

IP协议工作在网络层。IP协议提供了三个重要的定义。第一，IP协议定义了在整个TCP/IP互联网上数据传输所用的基本单元。因此，它规定了互联网上传输的数据的确切格式；第二，IP软件完成路由选择的功能，选择一个数据发送的路径；第三，IP包括了一组体现了不可靠数据传输的规则，这意味着不能保证数据一定能到达目的地。数据报可能被损坏，打乱顺序，或丢失。IP是TCP/IP互联网中最基本的部分，因此有时也称TCP/IP互联网为基于网际协议的技术。

◇ 网际控制报文协议（ICMP）

网际控制报文协议(ICMP)用于网络错误报告及产生要求注意的消息。ICMP报告的错误通常与数据报处理有关。ICMP只报告与数据报第0分片有关的错误。IP、UDP或者TCP层通常都在ICMP报文的基础上工作。ICMP一般属于TCP/IP中的IP层，但是依赖网络层的IP协议。ICMP报文被封装在IP数据报里面。

TCP/IP技术的发展与整个互联网的发展密切相关。在整个互联网上有数以万计的站点，这些站点上有着数百万的用户，互联网已经成为这些用户日常工作生活的一部分。我们已经度过了那种每个用户同时也是专家的早期开发阶段，进入了一个新的阶段，即只有很少的用户明白其中的技术。然而，不管表面上怎样，实际上无论是互联网还是TCP/IP协议都不是固定不变的。研究人员不断解决新的网络问题，工程师们也在不断改进底层机制。总之，TCP/IP技术在不断发展。

Fast Reading Two Windows Network Projector[①] Overview

A Windows Network Projector is a display device, such as a conference room projector, that uses Remote Desktop Protocol over an IP network to display the desktop of a Windows Vista-based PC. Windows Embedded CE 6.0 comes with an OS design template that allows you to create Windows Network Projectors.

A Windows Network Projector built with CE 6.0 and later is focused on supporting business scenarios such as those in the following list.

* Microsoft PowerPoint presentations with simple animations and still image display.

* Displays to a single projector (one-to-one connection).

* Mirror or extended display.

The Network Projector utilizes Remote Desktop Protocol (RDP) for display capabilities. It can support wired or wireless networks connections between the computer and the projector. It can support both ad hoc and infrastructure mode for wireless networks.

The Windows Embedded CE technologies behind Windows Network Projectors can allow you to build a number of different device types.

The following diagram (Pic 3.9) shows a direct implementation of Windows Network Projector built into a new or existing projector design. With this integrated support, the projector provides the capability of being discovered and connected to by a Windows Vista-based PC. This example shows the Windows Network Projector used with an infrastructure network connection.

Pic 3.10 shows the projector used with an ad hoc network connection.

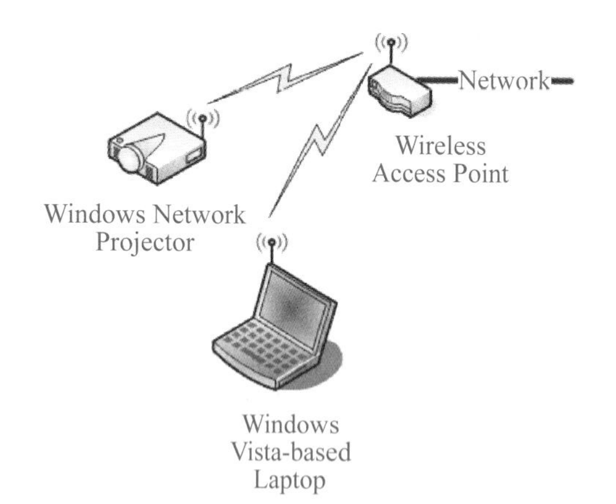

Pic 3.9 The projector built into a new or existing projector

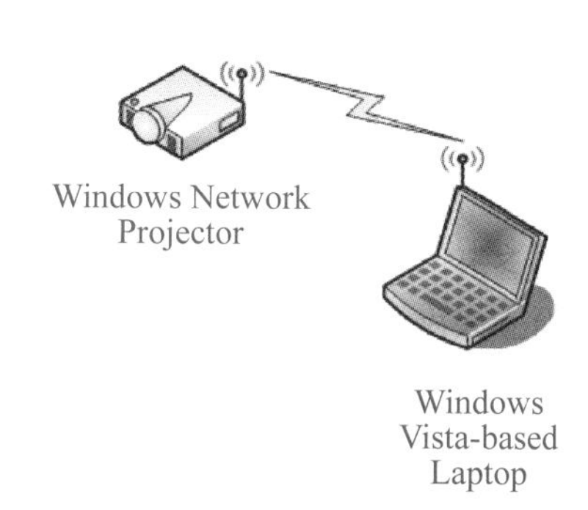

Pic 3.10 The projector used with an ad hoc network connection

① 资料来源于http://msdn.microsoft.com/en-us/library/ee480296

参考译文 Windows网络投影仪概述

Windows网络投影仪是一个显示设备，如会议室投影仪，使用IP网络中的远程桌面协议来显示装有Windows Vista系统的个人计算机的桌面。Windows Embedded CE 6.0内含操作系统平台，允许您创建Windows网络投影仪。

Windows网络投影仪内置CE 6.0，后来主要用于支持业务模式，具体内容如下：

*微软PowerPoint演示：有简单的动画，但仍是静态图像显示；

*单个投影仪展示（一对一的连接）；

*镜像或扩展显示。

网络投影仪使用远程桌面协议（RDP）显示功能，支持由有线或无线网络连接的电脑和投影仪，并同时支持无线自组网和无线基础设施模式。

Windows网络投影仪背后的Windows Embedded CE技术可以允许您构建许多不同的设备类型。

图3.9直观地展现了Windows网络投影仪与新的或现有的投影仪连用的设计。在这种综合保障下，投影仪就可以被装有Windows Vista系统的电脑发现，并连接到电脑上。该例子说明的是Windows网络投影仪与基础网络结构连用的情况。

图3.10则显示了投影仪可以用于自组网连接。

Exercise

Ex 1 **How many types of network are there? What are they? Can you give some examples about them?**

Ex 2 **Fill in the table below by matching the corresponding Chinese or English equivalents.**

Ethernet	
	客户/服务器
FDDI	
	对等网络
topology	
	电缆
Mesh network	
	城域网
Ring network	

Ex 3 Choose the best answer to the following questions according to the text we learnt.

1. Each computer or shared device found on the network is known as a __________.
   A. server
   B. workstation
   C. network node
   D. bridge
2. A __________ is a special computer on a network that provides & controls services (resources) for other computers on the network to use.
   A. workstation
   B. server
   C. bridge
   D. switch
3. __________ have limited access to the resources found on the network.
   A. Clients
   B. Administrators
   C. Nodes
   D. Servers
4. Every network has a "shape" which is normally referred to its __________.
   A. Bus
   B. topology
   C. Star
   D. Tree
5. In a __________ topology, each device has its own cable running to connect the device to a common hub or concentrator.
   A. Bus
   B. Mesh
   C. Star
   D. Tree

Part B Practical Learning

Training Target

In this part, students must finish two special tasks in English environment, under the guidance of the Specialized English teacher and the teaching related to Network Connecting in the computer laboratory. The students must work with each other in the same group.

Task One Connect the Computer to the Internet

In this task, students must connect their computers to the Internet through the campus network. The students can use cable to connect their computers to the laboratory's hub.

Task Two Construct LAN in English Environment①

In this task, students must construct a Local Area Network in English environment through the campus network.

There is some information about the construction of a LAN. The information will help the students accomplish the task.

To build a Local Area Network, you need to set both of the computers to have an IP address in the same subnet. For example, you can set one of your computer's IP address to be 192.168.0.1 and the next one as 192.168.0.2.

Make sure you use the subnet mask 255.255.255.0. By using this subnet mask, you can use any address from 192.168.67.1 to 192.168.67.254.

You can look at the example (Pic 3.11) for a visualization to see how the TCP/IP settings should be set up.

① 资料来源于http://www.makeuseof.com/tag/build-local-area-network-router/

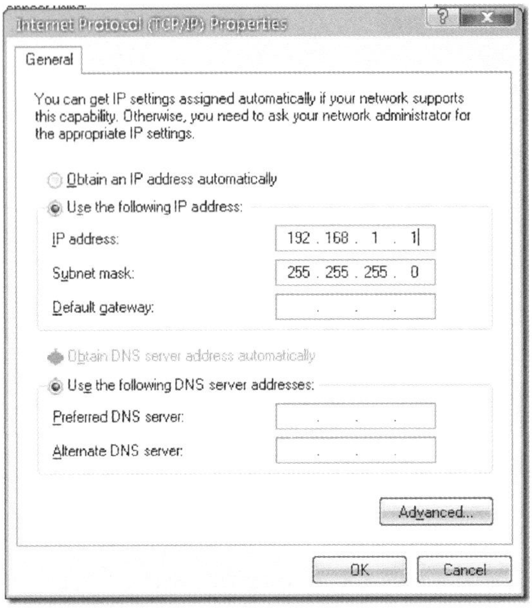

Internet Protocol (TCP/IP) Properties
General
You can get IP settings assigned automatically if your network supports this capability. Otherwise, you need to ask your network administrator for the appropriate IP settings.
Obtain an IP address automatically
Use the following IP address:
IP address: 192 . 168 . 1 . 1
Subnet mask: 255 . 255 . 255 . 0
Default gateway:
Obtain DNS server address automatically
Use the following DNS server addresses:
Preferred DNS server:
Alternate DNS server:
Advanced...
OK Cancel

Pic 3.11

The second computer should be set up like Pic 3.12:

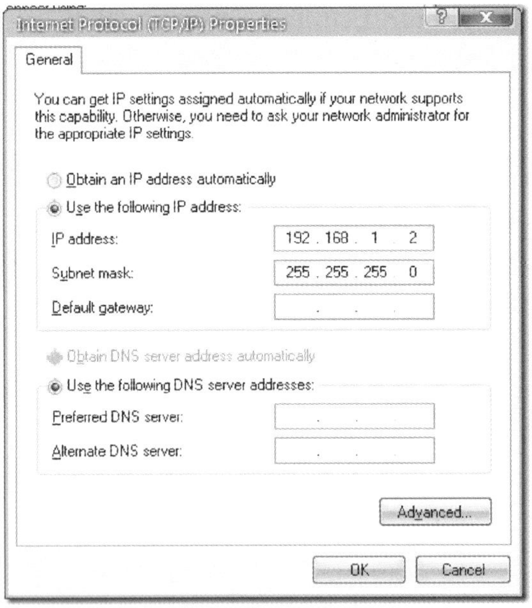

Internet Protocol (TCP/IP) Properties
General
You can get IP settings assigned automatically if your network supports this capability. Otherwise, you need to ask your network administrator for the appropriate IP settings.
Obtain an IP address automatically
Use the following IP address:
IP address: 192 . 168 . 1 . 2
Subnet mask: 255 . 255 . 255 . 0
Default gateway:
Obtain DNS server address automatically
Use the following DNS server addresses:
Preferred DNS server:
Alternate DNS server:
Advanced...
OK Cancel

Pic 3.12

By keeping both machines in the same subnet, they can "talk" to each other. In the example above, you can use 192.168.1.1 through to 192.168.1.254 if you keep the subnet mask the same at 255.255.255.0.

Once this is set up, you can share resources on both machines and be able to access them by launching the Run window and entering the other computer's IP address in this format \\192.168.1.1, like Pic 3.13:

Now, if you get a window that looks like Pic 3.14 instead of Pic 3.13, you need to set up your folder's security. You can do this by right-clicking on the folder you're trying to share/access and choose Sharing and Security. From here, you can give the appropriate permissions.

Pic 3.13

Pic 3.14

Part C Occupation English

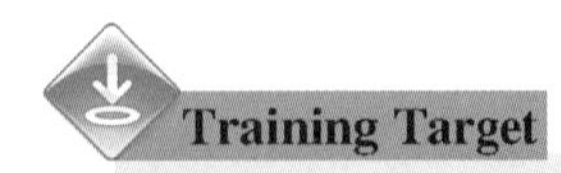

In this part, students are supposed to practice the dialogue to be skilled at troubleshooting.

Network Technical Support 网络技术支持

Post: Staff of General Computers Technical Support Center

(岗位: 通用电脑技术支持中心职员)

A : General Computers Technical Support Center. Can I help you?
通用电脑技术支持中心。我能帮助你吗?

B : I have opened an Online Purchase account on your website. The ads look good, but I still have my doubts. How secure can my account information be? Can I fully trust your security system?
我已经在你们的网站上开通了一个在线购物的账户。广告看起来很不错,但我仍然有一些顾虑。我的账户信息到底有多安全?你们的安全系统是否值得信赖?

A : I can assure you that General Computers Corporation provides the most up-to-date encryption technology for online payment. We use multiple certifying and encryption methods at the same time. Besides, we have an experienced expert team, who can respond immediately when there is a technical emergency.
我可以向你保证,通用电脑公司为网上支付提供了目前最先进的加密技术。同时我们还使用多重的验证和加密方法。另外,我们还有一支经验丰富的专家团队,他们可以对任何一个技术上的突发事件做出及时的处理。

B : Hmm, that sounds reassuring. Thank you for your explanation.
嗯,听起来很可靠。谢谢你的解释。

A : You're welcome. Thank you for using our products.
不客气。感谢你使用我们的产品。

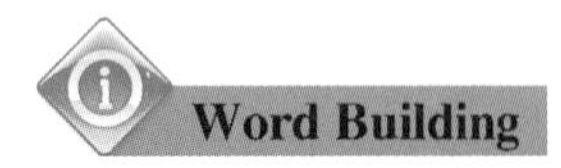

前缀/后缀由一个或几个字母组成,放在词根或单词之前/之后,组成一个新词。

(1) -ure (后缀): 行为、行为的结果、状态

fail: 失败 ———————— failure: 失败

press: 压、按 ———————— pressure: 压力

(2) -or（后缀）：……的人

visit：参观，访问 ———— visitor：来访者

invent：发明 ———— inventor：发明者

(3) -ous（后缀）：有……的

danger：危险 ———— dangerous：危险的

fame：名声 ———— famous：著名

(4) -ful（后缀）：充满……的，具有……性质的

use：用处 ———— useful：有用的

Ex Translate the following words and try your best to guess the meaning of the word on the right according to the clues given on the left.

expand	扩展（动词）	expansion ________
geometry	几何学（名词）	geometrical ________
repeat	重复（动词）	repeater ________
route	路（名词）	router ________
power	能力（名词）	powerful ________
flex	弯曲（名词）	flexure ________
essence	本质、精华（名词）	essential ________
magnetism	磁力（名词）	electromagnetism ________

Exercise

Ex 1 How many network devices do you know? Can you summarize the characteristics of them?

Ex 2 Fill in the table below by matching the corresponding Chinese or English equivalents.

twisted pair	
	交换机
repeater	
	路由器
optical fiber	
	集线器
OSI	
	网卡
UTP	
	物理地址

Ex 3 Choose the best answer to the following questions according to the text we learnt.

1. NICs are considered Layer 2 devices because each individual NIC throughout the world carries a unique code, called ________.
   A. binary impulses
   B. serial format
   C. code of network interface card
   D. Media Access Control Address
2. One of the disadvantages of the type of cable that we primarily use, CAT 5 UTP , is ________.
   A. the cost of this kind of medium
   B. difficult to install
   C. cable length
   D. too complicated
3. Consequently, bridges are concerned only with passing or not passing frames, based on ________.
   A. Internet Protocol address
   B. logical address
   C. their destination MAC address
   D. the format of the frame
4. The purpose of a router is to examine incoming packets, choose the best path for them through the network, and ________.
   A. then let them wait there for some time
   B. then change the information coming from the application layer
   C. then switch them to the proper outgoing ports
   D. then put them to another router
5. ________ is the best transfer media at present.
   A. Optical fiber
   B. Coaxial cable
   C. Twisted pair
   D. Radio
6. ________ can select the best path to route a message.
   A. NIC
   B. Bridge
   C. Switch
   D. Router

Project Four

Designing Online Store

Part A Theoretical Learning

Part B Practical Learning

Part C Occupation English

Part A Theoretical Learning

Training Target

In this part, our target is to improve the speed of reading professional articles and the comprehension ability of the reader. We have marked specialized vocabulary key words in some paragraphs so that the reader can quickly grasp the main idea of the sentences and paragraphs.

Skill One A Short Introdution to the Internet

Internet is a giant global and open computer network, which is a collection of interconnected networks. It is also a way to exchange information and share resource between **international** computers.

The original Internet is ARPANET(Advanced Research Projects Agency Network) , which was established by U.S. Department of Defense in 1969. The ARPANET was used to share data between some separated **military** institutes and universities in region. By 1972 the network had expanded to incorporate 40 **nodes**. Many U.S. government agency networks had been linked by ARPANET, and because the networks were of a disparate nature, a common network **protocol** called TCP/IP (Transmission Control Protocol&Internet Protocol) was developed and became the standard for inter networking military computers.

When the U.S. was developing their national net, the other countries were developing, too. At the end of 1980s, interlinkage of different countries' computer net appears. After that there were countries joining in every year and getting to form the present Internet. It is growing so quickly that nobody can say exactly how many users are "On the Net" now.

Once your computer makes a connection with the Internet, you will find that you have walked into the largest **repository** of information—a magic world. The following are the important service functions that the Internet provides.

1.E-mail

The most widely used tool on the Internet is electronic mail or E-mail. E-mail enables you to send messages to Russian, Japan and so on, no matter how far between individuals. E-mail messages are

interconnect [intəkə'nekt] v.互联

international [intə'næʃənəl] adj. 国际的

military ['militəri] adj. 军事的，军用的

node [nəud] n. 节点

protocol ['prəutəkɔl] n. 协议

repository [ri'pɔzitəri] n. 仓库，资源丰富的地方

generally sent from and received by **mail servers**—computers that are **dedicated to** processing and directing E-mail. As a very convenient and inexpensive way to transmit messages, E-mail has dramatically affected scientific, personal and business communications. In some cases, E-mail has replaced the telephone for conveying messages.

mail server 邮件服务器
dedicate to 用作……，奉献给

2.File Transfer Protocol

File Transfer Protocol (FTP) is a method of transferring files from one computer to another over the Internet, even if each computer has a different operating system or storage format. FTP is designed to **download** files (e.g. receive from the Internet) or **upload** files (e.g. send to the Internet). The ability to upload and download files on it is one of the most valuable features the Internet offers. This is especially helpful for those people who rely on computers for various purposes and who may need software drivers and **upgrades** immediately. Network administrators can rarely wait even a few days to get the necessary drivers that enable their network severs to function again. The Internet can provide these files immediately.

download [daun'ləud] v. 下载
upload [ʌp'ləud] v. 上传
upgrade [ʌp'greid] n. 升级，软件升级

3.The World Wide Web

The World Wide Web (WWW), which Hypertext Transfer Protocol (HTTP) works with, is the fastest growing and most widely-used part of the Internet. The WWW is a way to exchange information between computers and the Internet. **Hyperlink** makes the Internet easy to navigate. It is an object (word, phrase, or picture) on a webpage that, when clicked, transfers you to a new webpage. One of the main reasons for the extraordinary growth of the web is the ease in which it allows access to information. One limitation of HTTP is that you can only use it to download files, you can not upload them.

hypertext ['haipətekst] n. 超文本
hyperlink ['haipəliŋk] n. 超链接

4.Telnet

Telnet allows an Internet user to connect to a distance computer and use that computer as if he or she was using it directly. To make a connection with a Telnet client, you must select a connection option: "Host Name" and "**Terminal** Type". The host name is the IP address (DNS) of the remote computer to which you connect. The terminal type describes the type of terminal **emulation** that you want the computer to perform.

telnet ['telnet] n. 远程登录
terminal ['tə:minəl] n. 终端
emulation [,emju'leiʃən] n. 竞争，仿真

The Internet has many new technologies, such as global chat, **video conferencing**, free international phone and more. The Internet becomes more and more popular in society in recently years. So we can say that the Internet is your PC's window to the rest of the world.

video conferencing 电视会议

.End.

node n.节点
upload v.上传
host n.主机
hyperlink n.超链接
E-mail n.电子邮件
protocol n.协议
upgrade n.升级，软件升级
hypertext n.超文本
terminal n.终端
telnet n.远程登录
TCP/IP(Transmission Control Protocol/Internet Protocol) 传输控制协议和网际协议
WWW (World Wide Web) 万维网
FTP (File Transfer Protocol) 文件传输协议
HTTP (Hypertext Transfer Protocol) 超文本传输协议
DNS (Domain Name Server) 域名服务器
video conferencing 电视会议

参考译文 技能1 互联网简介

互联网是一个巨大的全球性开放式计算机网络，它由众多网络互联而成。它也是全球计算机之间实现国际间信息交流和资源共享的一种方式。

互联网的前身是美国国防部于1969年建立的ARPANET（高级项目研究机构网络），它是为了让在地域上相互分离的一些军事研究机构和大学之间实现数据共享而开发的。到1972年，这个网络扩展到可以包含40个节点。美国许多政府的网络也连入了ARPANET，由于网络的不同，一个被称为TCP/IP(传输控制协议和网际协议)的网络协议被发展起来，成为网络中的军用计算机都要遵守的标准协议。

在美国发展自己的网络的同时，其他国家也在发展自己的网络。到20世纪80年代末，不同国家的计算机网络连接出现了。从那以后，每年都有一些国家加入其中，逐步形成了现在的互联网。互联网发展如此之快以至于没有人能准确地说出网上到底有多少用户。

一旦你的计算机连入了互联网，你就会发现你进入了一个最大的信息宝库，一个神奇的世界。下面是互联网提供的重要服务功能。

1.电子邮件

在互联网上应用最广泛的是电子邮件，简称E-mail。通过电子邮件，你可以发送信息到俄罗斯、日本等，无论两人之间的距离有多远。通常都是由邮件服务器负责发送和接收邮件信息，邮件服务器是用来处理和传送邮件的计算机。利用电子邮件传递信息是非常方便而且便宜的方式，电子邮件已经极大地影响了科学、个人以及企业之间的通信。有些时候，电子邮件在信息传递上已经替代了电话。

2. 文件传输协议

文件传输协议（FTP）是一台计算机和另一台计算机之间通过互联网进行文件传输的一种方法，即使各计算机有着不同的操作系统和存储格式。FTP可以用来下载文件（例如从互联网上获取）和上传文件（例如发送给互联网）。上传和下载文件的功能是互联网所提供的最有价值的功能之一。这些功能（上传和下载）对于那些经常利用计算机实现各种应用，以及常常需要各类软件驱动程序且随时需要升级软件的人来说非常有帮助。当网络管理者急需要一些驱动程序来使网络服务器重新启动时，哪怕是几天的时间他们也是等不了的。通过使用FTP，互联网能够迅速提供这些文件。

3. 万维网

利用超文本传输协议（HTTP）的万维网（WWW）是互联网中发展最快、应用最广的部分。万维网是实现计算机与互联网之间信息传递的一种手段。超链接使我们漫游互联网更加容易。超链接指的是网页上的一些文字、短语、图片等，当单击它时，它会带你进入一个新的界面。网络飞速发展的主要原因是它很容易获取信息。HTTP的一个局限性是它只能用来下载文件而不能上传文件。

4. 远程登录

远程登录使互联网的用户能够登录到远程的计算机上，而且就像他（或她）亲自在使用那台计算机一样。要想登录到远程计算机上，你必须对主机名和终端类型进行选择。主机名就是你所要连接的计算机的IP地址（域名服务器），终端类型指的是你所希望计算机扮演的终端仿真的类型。

互联网还包含许多新的技术，如全球聊天、电视会议、免费国际电话等等。互联网近几年在社会上的影响越来越大，所以我们可以说，互联网是你的PC机通向世界其他地方的窗口。

Skill Two Website Design

Many people wish to create a flashy **website**. But creating a great website doesn't happen at the tips of the fingers, it happens in the depths of the brain. Outstanding websites result from extensive planning. **Prior** preparation saves time and avoids **frustration** both during page creation and when updates and additions are required. The three-step design **tutorial** will show you how to create a high-end attractive website.

Step One—Determining who use your site and their information needs

Successful websites know who their customers are and why they visit, and they provide a responsive and attractive display to those

website ['websait] n. 网站

prior ['praiə] adj. 在先的，优先的

frustration [frʌ'streiʃn] n. 挫折，挫败

tutorial [tju'tɔriəl] n. 辅导，指导

viewers. Customers don't visit our site because we spend time creating it; customers deserve maximum benefit from the time they **allocate** to us.

allocate ['æləkeit] v. 分配

Step Two—Editing your webpage

webpage ['webpeidʒ] n. 网页

1. Establish an **identity** and use it consistently on all pages

identity [ai'dentiti] n. 特性，图标

Viewers of our webpages should know exactly who we are, and after linking should know if they're still on one of our site's pages. It doesn't mean every page looks the same, but the colors and graphics we use should be consistent throughout the website. Establish a **theme** or identifying characteristic for your website.

theme [θi:m] n. 主题

2. Create user-friendly **navigation**

navigation [nævi'geiʃən] n. 导航

On a well-planned website, it's quick and easy to get to information pages—that's navigation. Plan navigation before pages are created. Establish a navigation plan to ensure that viewers quickly get what they need and that new pages of content can be quickly inserted and located.

3. Page **layout**

layout ['leiaut] n. 布置，规划

What's the difference between a webpage and a GREAT webpage? A webpage gives us information; a GREAT webpage catches our attention—that "Oh, Wow!" reaction—and gives information as expeditiously as possible—the key is planning and creativity.

To begin layout, analyze the information to be displayed and decide how it will be most readable. Pick the **template** that best accommodates that display. As your templates were created, page layout may have been anticipated. There are three methods to create balanced page layout: blockquote margins, tables, and **frames**. Each method has **pros and cons**; it can be advantageous to use all three to build a website.

template ['templeit] n. 模板，样板

frame [freim] n. 框架

pros and cons 优缺点

4. **Focus on** text

focus on 注意，以……为重点

Viewers come to website for information and if they don't get what they need, flash and glitz won't bring them back. The best websites pack essential information into well-organized and well-written text. Webpages should not, however, be too heavily written text. Surveys show that web users will not read long paragraphs of information. They prefer **concise**, bite-sized sections, clearly **delineated** so they can scan for the information they need. You should write essential content as clearly and concisely as possible with brief topic headers.

concise [kən'sais] adj. 简洁的，简明的

delineate [di'linieit] v. 叙述，描写

5. Use graphic images to enhance, not overpower

Graphics are a special challenge for web designers, requiring balance between overuse and **skimpiness**. A site filled with graphic images can have charm and impact. The secret for effective graphics is to stick to the theme and identity of the website.

skimpiness ['skimpinis] n. 简洁，节俭

Step three—Putting your new site on the web

1. **Domain name** registration

domain [dəu'mein] name 域名

The domain name is the address that users type into their web browsers (Internet Explorer or Netscape) to view your website. You select one or more proposed domain names such as Amazon.com or Buy.com. Your domain name must not be used by anyone else, and shorter is better. Choose a domain name that best reflects your business, products or services.

2. Publish the website

In this step you will be given instructions for uploading your new website to a computer known as a "server" where it makes your information available to any web surfer world wide. You will need an FTP client in order to upload files to your web server. CuteFTP (www.cuteftp.com) is highly recommended.

Once a Host has been selected, we will "publish" your new website for accessibility to everyone on the World Wide Web.

3. **Promote** your website

promote [prə'məut] v. 宣传，推广

Website promotion involves submitting your site address and search words to the top 12 search engines, which are Netscape, Yahoo, Microsoft, Alta Vista, Ask Jeeves, AOL, Excite, Google, Goto, HotBot, Looksmart and Lycos. These engines represent over 98% of all U.S. web searches. We can create special "headers" on your website that include all the necessary search engine friendly information. Other website promotion you may establish is: placing your web address on all stationery, business cards, and all broadcast and printed advertising media. You can begin as soon as your domain name is registered.

Creating and publishing your website is just "Tip of the Iceberg". The remaining 90% is unseen below the surface. These unseen features are important to the success of your website. They include **meta tag** creation, regular search engine submission, marketing exchange programs and many others.

meta tag 中继标签

.End.

Key Words

website n. 网站	layout n. 布置，规划
update v. 更新	template n. 模板，样板
webpage n. 网页	frame n. 框架
identity n. 特性，图标	theme n. 主题
domain name 域名	navigation n. 导航
promote v. 宣传，推广	meta tag 中继标签

参考译文 技能2 网站设计

许多人都希望能够创建一个吸引人的网站，但是创建一个好的网站并不是靠手指的敲击，而是要利用大脑进行设计。优秀的网站来源于全方位的设计，事先做好充分的准备既节省时间又可以避免在网页设计和更新的过程中遇到阻碍。下面的网站设计三步曲将指导你创建一个高水准的有吸引力的网站。

第一步——弄清谁使用你的网站及他们所需要的信息

好的网站都十分清楚它们的客户群是谁，为什么要访问网站，因此网站会为访问者提供一个及时的、有吸引力的界面。客户访问我们的网站不是因为我们花了时间去创建它，而是因为客户想通过访问我们的网站来获得最大的收益。

第二步——网页设计

1．创建图标并将此图标应用到所有的网页上

网站的访问者应该准确地知道我们是谁，点击相关链接后应该知道他们是否还在我们的网页上。这并不意味着每个网页看上去都一样，但是在整个网站设计中，颜色和图片应保持一致。要为你的网站创建一个主题或有标识性的特征。

2．创建界面友好的导航

在一个好的网站里获取信息是很快、很容易的，因为那里有导航。在网页创建之前要设计好导航，建立一个导航能保证访问者快速地获取他们所需要的信息，而且新的内容也能很快地添加进去。

3．页面布局

网页与好的网页的差别是什么呢？网页给我们提供信息；好的网页能吸引我们注意，使我们有“哦！哇！”的反映，并能尽可能快速地提供信息。两者的区别关键在于网页的设计与创意。

开始布局时，要分析所要显示的信息，确定如何才能使它们被更好地阅读。选择最合适的模板，一旦模板选定了，页面布局也就差不多了。这里有三种方法来创建协调的页面布局：第一种是页边缘设计；第二种是表格；第三种是框架。每一种方法都有优缺点，最有利的方法是

综合使用这三种方法来创建网站。

4. 文本的设置

访问者浏览网站是为了获得信息，如果他们得不到他们所需要的信息，即使是动画或图片也不会让他们停留。好的网站能够很好地组织和撰写重要的信息。然而，网页也不能包含过多的文字。调查表明，网络用户不喜欢阅读长的段落，他们喜欢短小精悍的文章，以便能快速浏览得到所需的内容。所以应该将重要的内容尽可能写得简单明了，当然还要有一个简要的标题。

5. 利用图片增强效果，但不能过于花哨

对网站设计者来说，合理地利用图片是另外一个难点，既不能过多，又不能太少。附有图片的网站比较有吸引力和影响力。利用好图片的一个有效方法是使图片紧扣主题和网站的内容。

第三步——网站发布

1. 域名注册

域名就是用户要访问你的网站时在网络浏览器（Internet Explorer或Netscape）中输入的地址。你可以选择类似Amazon. com或Buy. com这种形式的名字作为你的域名。你的域名必须得是别人没有用过的，并且最好不要太长。选择最能反映你公司的业务、产品或服务的名字作为你的域名。

2. 网站发布

这步操作将指导你把新建的网站上传到服务器系统上，上传后就可以使全世界的浏览者访问到你的网站。你需要使用FTP客户端将文件上传到你的网络服务器上。CuteFTP（www. cuteftp. com）是非常值得推荐的上传工具。

主机选定后，你就可以向全世界公布你的网站了。

3. 网站的推广宣传

网站的宣传包括向12大著名的搜索引擎提交你的网址和关键字，这12个搜索引擎是Netscape、Yahoo、Microsoft、Alta Vista、Ask Jeeves、AOL、Excite、Google、Goto、HotBot、Looksmart和Lycos。这些搜索引擎代表着全美国网站搜索98%以上的份额。我们还可以为网站创建特定的标题，使其包含更方便大部分搜索引擎搜索的信息。其他的网站宣传方法有：将你的网址添加到信笺、商业名片、广播或广告媒体中。当你的域名注册完后，你就可以立即开始对网站进行推广与宣传了。

网站的创建和发布就像“冰山一角”一样。其他90%的工作都是在底下看不见的。对于一个成功的网站来说，这些看不见的工作也是非常重要的。这些工作包括创建中继标签，规划搜索引擎的信息提交，规范市场交易系统，以及其他的推广活动。

Fast Reading One The Advancement of the Computer

The use of the transistor in computers in the late 1950s marked the coming of the second-generation computers. The most notable change was that transistors replaced vacuum tubes. This meant that the advent of smaller, faster, more reliable and less expensive computers than that were possible with vacuum-tube machines. In addition, the second-generation computers were given auxiliary storage, sometimes called external or secondary storage. Data was stored outside the computer on either magnetic tapes or magnetic disks. Using magnetic tapes or magnetic disks for input and output operations increased the speed of the computer.

RAM capacities increased from 8,000 to 64,000 words in commercially available machines by the 1960's, with access times of 2 to 3 ms (milliseconds). These machines were very expensive to purchase or even to rent and were particularly expensive to operate because of the cost of expanding programming. Such computers were mostly found in large computer centers operated by industry, government, and private laboratories—staffed with many programmers and support personnel.

Late in the 1960s the integrated circuit, or IC, was introduced, making it possible for many transistors to be included on one silicon chip. Therefore, the computers became even smaller and cheaper while their memory capacities became larger. The microprocessor became a reality in the mid-1970s with the large-scale integrated (LSI) circuit. The earliest microcomputer, the Altair 8800, was developed in 1975 by Ed Roberts; this machine used the Intel microprocessor and had less than 1 kilobyte of memory.

In the 1980's, very-large-scale integrated (VLSI) circuit, in which hundreds of thousands of electronic components were etched into a single silicon chip, became more and more common. Many companies, some new to the computer field, introduced in the 1970s programmable minicomputers supplied with software packages. The "shrinking" trend continued with the introduction of personal computers(PCs), some of which are programmable machines small enough and inexpensive enough to be purchased and used by individuals.

By the late 1980s, some personal computers were run by microprocessors that, handling 32 bits of data at a time, could process about 4,000,000 instructions per second. Microprocessors equipped with read-only memory(ROM), which stores constantly-used, unchanging programs, now performed an increased number of process-control, testing, monitoring, and diagnosing functions, like automobile ignition systems, automobile-engine diagnosis, and production-line inspection duties.

From the integrated circuit to large-scale integration and to very-large-scale integration, this was the start of the microprocessor age. The microprocessor continued to improve from the 80286, 80386 to the 80486, then Pentium, Pentium Ⅱ and so on.

Modern digital computers are all conceptually similar, regardless of the size. They can be divided into several categories on the basis of cost and performance: the personal computer or microcomputer, a relatively low-cost machine, usually of desktop size. It also includes laptops which are small enough to fit in a briefcase and palmtops which can fit into a pocket; the workstation, a microcomputer with enhanced graphics and communications capabilities that make it especially useful for office work;

the minicomputer, generally too expensive for personal use, is suitable for a business, school, or laboratory; the mainframe computer, a large, expensive machine which meets the needs of major business enterprises, government departments, scientific research establishments; the supercomputer, the largest and fastest computer.

The "fifth-generation" computer is using new technologies in very large integration, along with new programming language, and will be capable of amazing feats in the area of artificial intelligence, voice recognition. One important parallel-processing approach is neural network, which mimics the architecture of the nervous system.

This picture (Pic 4.1) shows China's first petaflop/s scale supercomputer—Tianhe-1. The Chinese

Pic 4.1 Tianhe-1

National University of Defense Technology (NUDT) recently unveiled China's fastest supercomputer, also the world fifth fastest computer, which is able to do more than one petaflop calculations per second theoretically at its peak speed. The TH-1 is made up of 80 compute cabinets including 2560 compute nodes and 512 operation nodes. The TH-1 system will be used to provide high performance computing service for the Tianjin area and the northeast of China. NSCC (National Supercomputer Computer Center) -TJ plans to use this system to solve the computing problems in data processing for petroleum exploration and the simulation of large aircraft designs. Other uses for the TH-1 supercomputer include the sciences, financial, automotive and shipping industries.

One continuing trend in computer development is micro-miniaturization, the effort to compress more circuit elements into smaller and smaller chip space.

.End.

参考译文 计算机的发展

在20世纪50年代后期，晶体管的应用标志着第二代计算机的问世。其最显著的变化就是晶体管代替了电子管。使用晶体管可以制造出速度更快、性能更可靠、价格更便宜的计算机，与电子管计算机相比体积也更小。另外，第二代计算机具有辅助存储器（也可称为外存或辅存）。数据可以存储在计算机外的磁带或磁盘上。使用磁带或磁盘进行输入/输出操作可以提高计算机的运算速度。

到了20世纪60年代，商用机中RAM的容量从8000字增长到64000字，访问时间为2～3毫秒。购买这些机器的价格相当昂贵，连租用费用都很高，这是因为不断扩充的程序成本使得运行费用居高不下。这样的计算机主要应用于工业、政府和私人实验室的大型计算机中心，并由许多程序员和维护人员进行操作。

20世纪60年代晚期，集成电路的引用使得许多晶体管可以嵌在一个硅片内，因此计算机的体积更小，价格更便宜，但存储容量却更大了。20世纪70年代中期，大规模集成电路的微处理器问世了。最早的微型计算机Altair 8800于1975年由Ed Roberts制造，这台计算机采用了英特尔微处理器，并且有不到1000字节的内存。

20世纪80年代，超大规模集成电路使成百上千的电子元件集成在一个硅片内变得越来越普遍。许多公司，包括一些新涉足计算机领域的公司，都在20世纪70年代引进了由软件包支持的可编程小型计算机。随着个人计算机的产生，“收缩”的趋势仍在继续，这些可编程机器体积小，价格便宜，足以满足个人购买和使用的需求。

到了20世纪80年代后期，一些个人计算机由一次可处理32位的微处理器来运行，每秒大约可处理400万条指令。微处理器装配有只读存储器（ROM），用来存储经常使用的、不变的程序，这样的微处理器可实现越来越多的过程控制、测试、监视和诊断功能，譬如汽车点火系统、汽车引擎诊断和生产线检查等功能。

从集成电路到大规模集成电路再到超大规模集成电路，揭示着微处理器时代的开始。微处理器的型号也由80286、80386、80486一直改进到Pentium系列。

现代数字计算机无论大小，其设计理念都基本相似。依据成本和性能，基本上可以分成几类：个人计算机（微型计算机），价格不高，通常指台式机，也包括可以装入公文包的便携式计算机和可以装入衣服口袋的掌上电脑；工作站，一种具有较强的绘图和通信能力的计算机，通常在办公室使用；小型计算机，一般比较昂贵，不适合个人使用，适用于公司、学校或图书馆；大型机，是一种大型的昂贵的计算机，主要适用于企业、政府部门、科研机构等；巨型机（超级计算机），是最大型、最昂贵的计算机。

第五代计算机将使用超大规模集成电路新技术和新程序语言，也将在人工智能、语音识别中有惊人业绩。还有一个重要的并行发展的分支是神经网络，它能够模仿人类神经网络的结构。

图4.1展示的是中国第一台千万亿次超级计算机——天河一号。它是由国防科技大学（NUDT）研制出来的中国速度最快的超级计算机，在世界上排名第五，它理论上能够达到每秒钟千万亿次的峰值速度。天河一号有80个机柜，包含2560个计算节点和512个服务节点。天河一号将为天津以及我国的东北地区提供高性能的服务。国家超级计算机天津中心计划将天河一号应用于石油勘探、航空飞船模拟设计、科研、金融、汽车制造、运输等方面的数据处理。

计算机的一个发展趋势是小型化，即将更多的电路元件压缩在更小的芯片上。

Fast Reading Two Storage Devices

We know that the CPU controlled by program can process data. Then where are the data and the program from? The answer is storage devices. We usually divide the storage devices into two types: the main memory and the secondary storage. A CPU can only execute the instructions of a program which has already been in the main memory.

The main memory of most computers is composed of RAM. A programmer can read and write RAM. We can store data and programs into RAM. When we have finished using them, we can let new ones occupy the position of the main memory, destroying the old ones. In a word, the content of RAM is easy to change. Sometimes we don't want the content of memory to be changed, for example, the automatic teller terminals used in many banks. They are controlled by a small computer, which is controlled by a program. If someone can modify the data, it may give free access to certain accounts, the bank would not allow such things to happen. In fact, these programs are stored in ROM, which we can only read but cannot modify. In a word, ROM is permanent memory that can be read, but not be written. How can a program or data enter the computer system? We often use diskette drive to copy them into the main memory. Then we come to the concept of secondary storage.

Hard disk

The hard disk is also called the hard drive, hard disk drive or fixed disk drive. The hard drive is the primary device that a computer uses to store data. Most computers have one hard drive located inside the computer case. If a computer has one hard drive, it is called "drive C ". If a computer has additional hard drives, it is called "drive D, E, F ", and so on. And the hard drive light is on when the computer is using the hard drive. Do not move the computer when this light is on.

The hard drive magnetically stores data on the stack of rotating disks, called platters. And a hard drive has several read/write heads that read and record data on the disks. A hard drive can store your programs and data files.

How shall we choose a hard drive? The first factor is the capacity. The amount of information a hard drive can store is measured in bytes. A hard drive with a capacity of 2GB to 20GB will suit most home and business users. Purchase the largest hard drive you can afford. A hard drive will be quickly filled up with new programs and data. For example, Microsoft Word is a word processing program that requires about 16 MB of hard drive space. The second factor is average access time. The average access time is the speed at which a hard drive finds data. It is measured in milliseconds (ms). One millisecond equals 1/1000 of a second. Most hard drives have an average access time of 9 to 14 ms. The lower the average access time is, the faster the hard drive will be. Up to now, there are several connection types of the hand disk, such as IDE, EIDE, SCSI and so on.

Removable hard disk

An interesting compromise between internal and external hard disks is the removable hard disk drive tray. A tray is installed into a standard PC case drive bay that allows regular internal hard disks to be placed into it. You can then swap the internal hard disk with another one without opening up the case, allowing you to use hard disks as a removable storage medium. In a way, the concept is similar to the way a removable car stereo is designed. These trays are also commonly called mobile racks or drive caddies.

For certain applications, this is the ideal removable storage device: it uses regular hard disks, which are very fast, highly reliable, very high capacity and very inexpensive. They can be used for backup purposes.

If you decide to use a mobile rack system, be sure to check out the specifications of the unit you are considering carefully. Different models come with support for different speed drives, some are made primarily of metal and others of plastic, and so on. Metal units will provide better cooling than plastic ones.

Videodisk

Videodisk is read and written by a laser beam. There is no physical contact between the recording surface and the read/write mechanism. Fast, accurate, compact, and easy to use, videodisk has become more and more popular.

.End.

参考译文 存储设备

我们知道CPU是在程序的控制下处理数据的。那么，数据和程序是从哪里来的呢？答案是存储设备。我们通常把存储设备划分成两类：主存储器和辅助存储器。CPU只能执行已经存放在主存中的程序指令。

大多数的计算机主存是由随机存取存储器(RAM)组成的。编辑器可读写RAM。我们可以把数据和程序存入RAM。当我们完成操作时，可以把新的内容复制到主存中。这个复制过程洗刷掉了旧内容。总之，RAM的内容容易被更改。有时我们不想改变存储器的内容。例如，用于很多银行中的自动取款机，它们由小型计算机控制，而小型计算机又由程序控制。如果有人能修改数据，便可以自由访问某些账户，银行是不允许这种事情发生的。事实上，这些程序只存放在只读存储器中，我们只能读取而不能更改。总而言之，ROM是永久性的存储器，只能读不能写。怎样能把程序和数据输入计算机系统呢？我们通常利用磁盘驱动器把它们拷贝到主存里，于是便引入了辅助存储器的概念。

硬盘

硬盘也叫作硬驱、硬盘驱动器或固定磁盘驱动器。硬盘驱动器是计算机用来存储数据的主要设备。大多数计算机都在机箱内设置一个硬盘驱动器。如果计算机只有一个硬盘驱动器，它就被称作“C盘”。如果计算机有另外的硬盘驱动器，则被称作D盘、E盘、F盘等。当计算机在使用硬盘驱动器时硬盘指示灯是亮的。当硬盘指示灯亮时不要移动计算机。

硬盘利用磁力将数据存放在旋转磁盘的轨道上，称为磁碟。硬盘上有几个用来存取数据读写的磁头。硬盘能存储程序和数据文件。

我们应该怎样选择硬盘呢？首要参数是容量。硬盘所能存储的信息数量是用字节来衡量的。2GB到20GB容量的硬盘能满足大多数的家庭和商业用户的需求。在你能负担的范围内，尽量购买大容量的硬盘。硬盘会很快被新的程序和数据占满。例如，微软Word是一个要求16MB硬盘空间的文字处理程序。第二个参数是平均存取时间，它是硬盘寻找数据的速度，以毫秒为单位。1毫秒等于1/1000秒。大部分硬盘的平均存取时间在9～14毫秒之间。平均存取时间越少，硬盘的速度就越快。现有的几种硬盘连接类型有IDE、EIDE、SCSI等。

移动硬盘

在内部和外部硬盘之间折中的方法就是使用可移动硬盘驱动器托盘。托盘安装在标准的PC机箱驱动器槽上，该托盘允许一般的内部硬盘安装在上面。然后内部硬盘与另一个硬盘便可以交换却不必打开机箱，从而允许你像使用移动存储介质一样地使用硬盘。在某种程度上，它的设计类似于一套可移动的汽车立体音响。这些托盘也叫做移动架或驱动盒。

对某些应用来说，移动硬盘是理想的可移动存储设备：它使用常规硬盘，速度快，可靠性高，容量大，并且价格便宜。移动硬盘还可用于备份。

如果你决定使用一个移动硬盘系统，一定要仔细地检验你所考虑的这种硬盘的规格。不同规格的硬盘支持的速度不同，有些主要是由金属构成的，还有一些则是由塑料制成的。金属的比塑料的更容易冷却。

光盘

光盘是通过激光束来读写的，记录表面和读写装置没有物理接触。快速、精确、紧密和易使用的光盘已经变得越来越流行。

Exercise

Ex 1 What do you do on the Internet? Can you give us some examples?

Ex 2 Fill in the table below by matching the corresponding Chinese or English equivalents.

protocol	
	主机
DNS	
	超文本传输协议
FTP	
	中继标签
video conferencing	
	远程登录
terminal	
	超链接

Ex 3 Choose the best answer to the following questions according to the text we learnt.

1. Which protocol does the Internet mainly use? ________

   A. The OSI reference model

   B. TCP/IP

   C. File Transfer Protocol

   D. HTTP

2. An example of a client-server application on the Internet would be________.

   A. NIC

   B. E-mail

   C. Word

   D. hard drive utilities

3. Which service dose the Internet not provide? ________

   A. E-mail

   B. WWW

   C. ASP

   D. FTP

4. A protocol is a set of _________ .

   A. drivers

   B. servers

   C. regulations

   D. hardwares

5. What is used to specify placement of text, files, and objects that are to be transferred from web server to web browser? _________

   A. HTTP

   B. HDLC

   C. HTML

   D. URL

Part B Practical Learning

Training Target

In this part, students must finish two special tasks in English environment, under the guidance of the Specialized English teacher and the teaching related to online store design in the computer laboratory. The students must work with each other in the same group.

Task One Collect the Necessary Data

In this task, students must collect necessary data for the online store.

The data includes the name of your store, the form of your store and the platform to open your store, choose the product, and be clear about the details of the goods and so on.

Task Two Design the Online Store[1]

In this task, students must design their online store. There is some information about designing online store. I hope the information can help students finish their task.

Four Things You Want to Have

1. Registered Domain
2. Premium Web Hosting Account with cPanel and auto-installer support
3. WordPress—an open source platform you need
4. List of your Fashion Products with their respective prices and descriptions

1. Registered Domain

To start, you need to have a professional aura, a sense of unique identity of your online store so that your potential online customers and other site visitors can locate your site. This is possible by getting a registered domain name for your site at Name.com and at any other accredited best domain registrar.

2. Premium Web Hosting Account with cPanel and auto-installer support

Steps in Signing Up on HostGator

(1) Browse through to HostGator.com and select your web hosting plan (Pic 4.2).

[1] 资料来源于http://www.startonlinestore.org/ 有改动

Pic 4.2

(2) Enter your domain name. Input any validated HostGator promotion code you have got (Pic 4.3).

Step 1: Choose a Domain   Step 1 » Step 2

Register a New Domain
Help me register a new domain name.
Enter Domain Name:
com
CONTINUE TO STEP 2

or

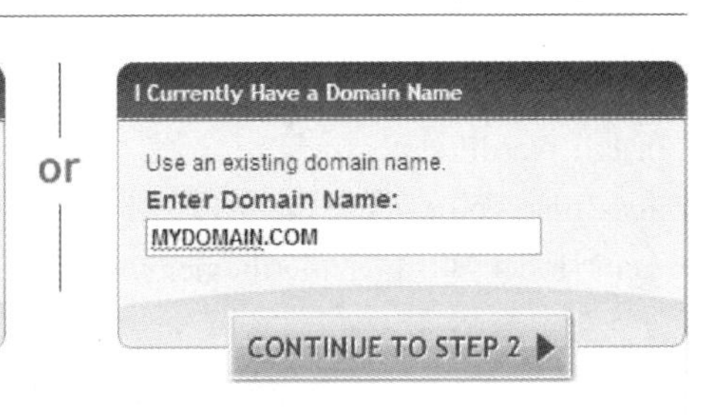

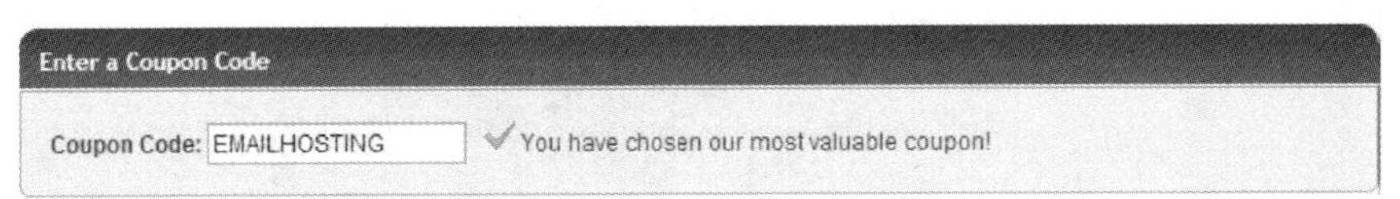

Pic 4.3

(3) Select Package Type and Billing Cycle (Pic 4.4).

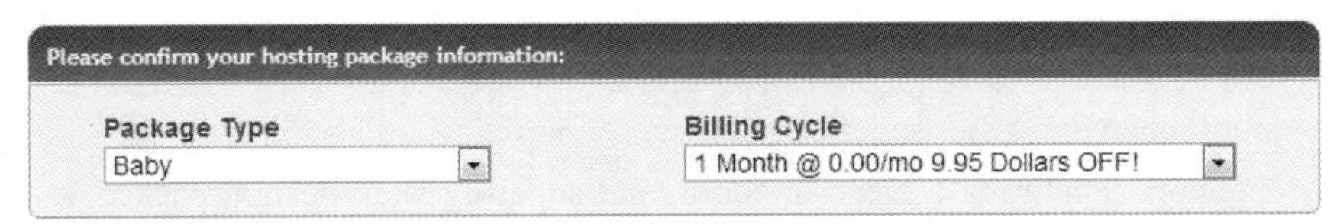

Pic 4.4

(4) Make your Account along with your billing specifics (Pic 4.5).

Please choose your account information:
Username
May not contain capital letters
Must start with a letter
Must be 2-8 characters long
May not contain special characters
Security Pin
Must be 4-8 characters long
May only contain numbers

Pic 4.5

(5) Finally, create your account (Pic 4.6).

Please review the order details below:

24/7/365 Phone, Live Chat, Email Support	FREE!
Instant Account Activation	FREE!
Money Back Guarantee!	45 Days
Hosting	Baby
Package (1 Month Cycle)	$9.95
Coupon Credit (EMAILHOSTING | 9.95 Dollars OFF!)	-$9.94
Hosting Addons	$0.00
Existing Domain (mydomain.com)	$0.00
Setup Price	$0.00
	Total Due: $0.01

☐ I have read and agree to the terms and conditions of use.

CREATE ACCOUNT

Pic 4.6

3. WordPress—an open source platform you need

First, Install WordPress.

After installing, you can use the WP.

(1) Mercor (Pic 4.7)—amazing theme, you'll love the sticky memo and the excellent design.

Pic 4.7

(2) Neighborhood (Pic 4.8)—super responsive, retina ready, and built upon the 1170px Twitter Bootstrap framework. Featuring a clean, modern, and superbly slick design, packed with the most powerful Swift Framework which offers limitless possibilities.

Pic 4.8

(3) Primashop (Pic 4.9)—future-proof WooCommerce compatible and it is compatible with most of WooCommerce extensions.

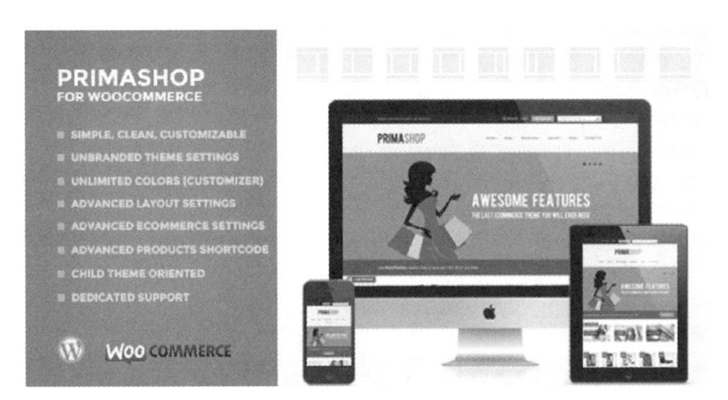

Pic 4.9

(4) Camp (Pic 4.10)—easy-to-customize and fully featured E-Commerce WordPress Theme.

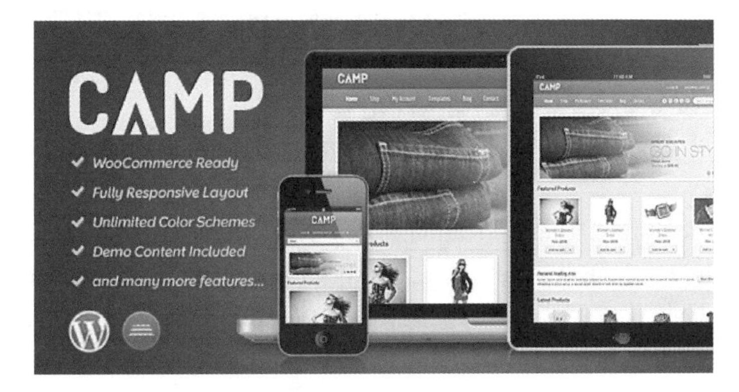

Pic 4.10

4. List of your Fashion Products with their respective prices and descriptions

After you install everything you need and customize what needs to be customized, it's time for you to add your products. Your listing becomes handy by adding items on your site. Most E-Commerce will add a menu tab for the items. So simply head into that menu, then place your product title, price, description and most importantly your product image. You have to do manually the adding of your own products. Add and edit your products as often as you need to.

Congratulations, you have successfully created your own online store!

Part C Occupation English

Training Target

In this part, students are supposed to practice the dialogue, and act as staff of network customer service center. Offering help to the customers, especially creating a new connection is one of the common tasks in their new job.

Creating a New Connection 创建新的网络连接

Post: Network Customer Service Staff （岗位：网络客服人员）

作为网络客服人员，需要向顾客提供各种有关上网的咨询服务。网络连接是常见的问题之一。

A : Hello. Is that Customer Service? Will you tell me how to log onto my ADSL network? How can I create an ADSL connection on my Windows XP?

你好！客服中心吗？请告诉我怎样登录ADSL网络？我该怎样在Windows XP操作系统上建立一个ADSL连接？

B : Oh, it's not difficult to make the creation if you have all the hardware in the right place. In Windows XP, in the Control Panel, open Network Connection. Double click the Create New Connection icon, and you'll see a New Connection Wizard. In the next page, choose Connect to the Internet, and then choose Connect using a broadband connection that requires a username and password or the one that is always on. You may create a new ADSL connection in this way. 你好！如果你的硬件都没有问题，建立连接不难。在Windows XP操作系统下，打开控制面板，进入网络连接。双击“创建新连接”图标，你会看到“新连接向导”。在下一页，选择“连接到互联网”，然后选择要求用用户名和密码的“宽带连接”来连接。这样你就可以创建一个新的ADSL网络连接了。

****

A : Good morning. I bought a Wireless LAN adapter, and I'm trying to use it on my Windows XP system. Will you give me some instruction so that I can access the network? Please tell me how to enable the connection.

早上好!我买了一个无线网卡，想把它装在Windows XP系统上。请你教我怎么安装才可以登录网络。告诉我怎样连接。

B : You should first check whether your operation system has recognized your WLAN adapter in your Device Manager. Then you should try to configure it in the correct way. Open Properties interface for TCP/IP in your adapter properties page, and let the system automatically obtain its IP and DNS address.

你首先要在“设备管理器”中检查你的操作系统是否识别无线网卡。然后你要用正确的方法安装。在网卡“属性”页面中，打开“TCP/IP属性”界面，让系统自动获取IP和域名服务器地址。

Word Building

前缀/后缀由一个或几个字母组成，放在词根或单词之前/之后，组成一个新词。

(1) inter-(前缀)：在……之间；中间

connect：联系——————interconnect：互联

act：行动，起作用——————interact：互相作用

(2) hyper-(前缀)：超出

text：文本——————hypertext：超文本

link：链接——————hyperlink：超链接

(3) un-（前缀）：反、不、非

format：格式化——————unformat：未格式化

delete：删除——————undelete：反删除，恢复删除

(4) -ness（后缀）：情况，性质

ill：有病的——————illness：疾病

idle：懒惰的——————idleness：惰性

Ex Translate the following words and try your best to guess the meaning of the word on the right according to the clues given on the left.

national	国家的（形容词）	international ______
media	媒体（名词）	hypermedia ______
install	安装（动词）	uninstall ______
repository	仓库（名词）	repost ______
unprecedented	前所未有的（形容词）	precedence ______
retailer	零售商（名词）	retail ______
notification	通知（名词）	notify ______
norm	标准（名词）	normal ______
skimp	节俭的（形容词）	skimpiness ______
graphic	图形的（形容词）	graph ______

Exercise

Ex 1 You have tried to create your own website. But once you have owned your website, how to make it perfect?

Ex 2 Fill in the table below by matching the corresponding Chinese or English equivalents.

layout	
	网站
homepage	
	网页
domain name	
	框架
navigation	
	模板
web address	

Ex 3 Choose the best answer to the following questions according to the text we learnt.

1. What does the word "frustration"mean in the first paragraph of *website Design*? ________
   A. Lacking of programming knowledge
   B. Setting up a website is very difficult
   C. Not having extensive planning
   D. Having no time to surf the Internet
2. Which of the following statements are not the steps of website design? ________
   A. Determining consumers　　B. Webpage design
   C. Surfing the World Wide Web　　D. Publishing the website
3. Which of the following statements are not the steps that you have done when editing your webpage? ________
   A. Creating user-friendly navigation
   B. Page layout
   C. Publishing the website
   D. Using graphics
4. When you select domain name for your website, what is the important idea ? ________
   A. Not be used by anyone else
   B. Best reflect your business or products
   C. Not too long
   D. All of the above
5. Which of the following is not the search engine ? ________
   A. Google　　B. Baidu
   C. Netscape　　D. Dreamweaver

Project Five

Beautifying the Online Store Page

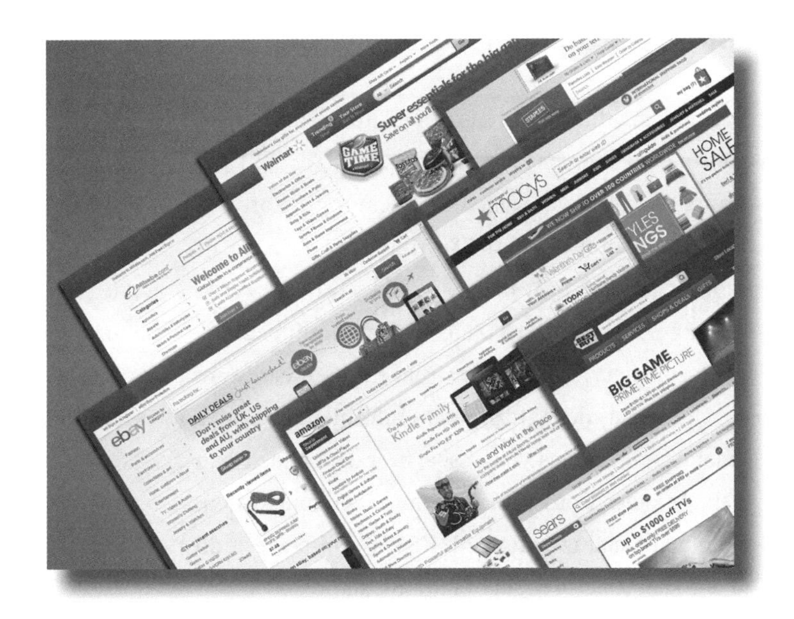

Part A Theoretical Learning

Part B Practical Learning

Part C Occupation English

Part A Theoretical Learning

Training Target

In this part, our target is to improve the speed of reading professional articles and the comprehension ability of the reader. We have marked specialized vocabulary key words in some paragraphs so that the reader can quickly grasp the main idea of the sentences and paragraphs.

Skill One A Short Introdution to Multimedia

The concept of multimedia has been around for years. If we break the word multimedia into its component parts, we will get multi—meaning more than one, and media—meaning a form of communication. So what is multimedia? **Multimedia is a kind of computer technology that combines text, audio, static graphic images, animations, and full-motion video.** In fact, multimedia is just two media: sound and pictures. It is made from a mix of hardware and software, or machine and ideas. It presents information, shares ideas and evokes emotions. It enables you to see, hear, and understand the thoughts of others. In other words, it is a form of communication. You can control multimedia and interact with it. You can make it do what you want it to do. That's the strength of multimedia.

audio ['ɔ:diəu] n. 音频
animation[æni'meiʃən] n.动画
video ['vidiəu] n. 视频

◇ **Audio**

Audio refers to sound or to things which can be heard. Usually, the human ear can hear a range of frequencies between around 20Hz~20KHz. But some animals can hear higher frequencies. Computer software can easily process digitized sound. Virtually, all professional sounds of recording and editing are digital nowadays.

◇ **JPEG(Joint Photographic Experts Group)**

JPEG image-compression standard and file format define a set of **compression** methods for high-quality images, such as photographs, single video **frames** or scanned pictures. JPEG does not work very well when compressing text, line art or vector graphics.

JPEG (Joint Photographic Experts Group) abbr.联合图像专家组
compression [kəm'preʃən] n. 压缩
frame[freim] n. 帧

◇ **MPEG(Motion Picture Experts Group)**

MPEG standards are the main **algorithms** used to compress video and have become international standard since 1993. Because movies contain both images and sound, MPEG can compress both video and audio. **A major goal of the MPEG-2 video standard is to define the format of the video data which is to be transmitted.** This data format is the result of a compression and encoding process. If the compression techniques are used in MPEG-2, then they are to a large extent based on some knowledge. The knowledge is that we have about how the human eye and the visual centers in the brain recognize images.

MPEG (Motion Picture Experts Group) abbr. 运动图像专家组
algorithm ['ælgəˌriðəm] n. 算法

◇ **Digital Video**

So far, the capabilities of the human eye to recognize images have been described. But how the images are described in digital equipment? Video applications deal with so called color spaces in order to define images. There are two major color spaces types used in digital video: RGB and YUV. RGB is commonly used in computer environments, while YUV is related more to the television world.

The simplest representation of digital video is a sequence of frames, each consisting of a **rectangular grid** of picture elements or **pixels**. Each pixel can be a single bit to represent either black or white, using 8 bits per pixel to represent 256 gray levels. Thus it gives high-quality black-and-white video. For color video, good systems use 8 bits for each of the RGB colors. 24 bits per pixel limits the number of colors to about 16 million, however, the human eye cannot even distinguish so many colors.

rectangular [rek'tæŋgjulə] adj. 矩形的
grid [grid] n. 网格
pixel['piksəl] n. 像素

.End.

Key Words

audio n.音频
video n.视频
frame n.帧
pixel n.像素
animation n.动画
compression n.压缩
algorithm n.算法
JPEG (Joint Photographic Experts Group) abbr.联合图像专家组
MPEG (Motion Picture Experts Group) abbr.运动图像专家组

参考译文 技能1 多媒体简介

多媒体的概念已经出现好多年了。如果我们把multimedia这个词分开，我们便得到multi——多，和media——通信的方式。那什么是多媒体？多媒体是一种将文本、音频、静态图像、动画和动态视频结合为一体的计算机技术。实际上多媒体只是两个媒体：声音和图像。它由硬件和软件，或者说机器和思想混合而成。它可以展示信息、交流思想和抒发情感。它让你看到、听到和理解其他人的思想。也就是说，它是一种通信的方式。你可以控制多媒体，可以与多媒体交互，可以让它按你的需要去做。这就是多媒体的优势。

◇ 音频

音频就是指语音或者能被人们听到的声音。通常人的耳朵可以听见的频率范围大约在20Hz到20KHz之间，而某些动物却能听到更高频率的声音。计算机软件可以很容易地处理数字化声音。今天，录制和编辑的专业化声音几乎都是数字化的。

◇ JPEG（联合图像专家组）

JPEG图像压缩标准和文件格式为高品质图像定义了一套压缩方法，适用于照片、单帧视频信息和扫描图像等。JPEG在处理文本、线条图或矢量图形时效果不是很好。

◇ MPEG（运动图像专家组）

MPEG标准是用来压缩视频的主要方法，自1993年以来已成为国际标准。电影包括图像和声音两个部分，MPEG既可以压缩视频，又可以压缩音频。MPEG-2视频标准的主要目的是定义要传输的视频数据的格式。该数据格式是经压缩和编码处理后的结果。用于MPEG-2中的压缩技术在很大程度上是以目前的一些知识为基础的。这些知识是我们所拥有的、有关人眼与大脑视觉中心识别图像原理的知识。

◇ 数字视频

迄今为止，人们已经能描述人眼识别图像的能力。但在数字化设备中图像是如何被描述的呢？视频应用软件处理用所谓的色彩空间来定义图像。在数字视频中，主要使用两类色彩空间：RGB和YUV。RGB常用于计算机环境，而YUV则更多地与电视领域相关。

最简单的数字视频表示法是帧序列，每帧包含一个矩形的图形元素（像素）网格。每个像素是一个简单的黑点或白点。每个像素用8比特来表示256个灰度级。这样便能产生高质量的黑白视频。对于彩色视频，好的系统对每个RGB颜色各用8比特来表示，而每像素24比特时颜色可达1600万种，人眼完全无法辨别如此多的颜色。

Skill Two Multimedia Devices

A multimedia PC is used in various aspects. Its **appearance** is a revolution in the computer field. Multimedia PC needs to be more powerful than mainstream computer. At least the multimedia PC defines the mainstream. Multimedia is the combination of sound, graphics, animation, video and text. It makes the computer play an important role in developing applications. **Multimedia PCs can run multimedia applications**, normally equipped with a sound card, a CD-ROM drive and a **high-resolution** color monitor. **In addition**, there are also scanners, digital cameras and so on. An ordinary computer differs greatly from multimedia, the only things are a soundboard and a CD-ROM driver. The CD serves as multimedia's chief storage and exchange medium.

appearance [ə'piərəns] n. 出现

high-resolution 高分辨率

in addition 此外

CPU	Intel i7 8700K CPU 8 generations of six cores
Motherboard	GIGABYTE Z370 AORUS Gaming 5
Memory	Kingston DDR4 2133 16GB
HDD	SAMSUNG 850 EVO 1TB SATA3
Graphics	Galaxy GTX 1080 GAMER 1683(1822) MHz/10000MHz 8G/256Bit
Display	SAMSUNG U28E590D 28 inch 4K High-definition image LED Backlight
CD-ROM	LG 24 double speed SATA Interface Build-in DVD
Case	Golden field Extraordinary Z2
Power	CoolerMaster 550W V550
Speaker	HiVi M200MKIII+ HIFI 2.0 Bluetooth

◇ **Sound card**

The sound card is an add-on device that generates analog sound signals **from** digital data, using either a **digital-to-analog converter** or an FM synthesis chip. There are three major standards for PC sound cards: AdLib, Sound Blaster, and Windows-compatible. Many PC sound cards include electronics to generate sounds from MIDI data on-board. There are two kinds of MIDI sound generations: **FM synthesis** simulates musical notes by modulating the frequency of a base carrier wave; whereas **waveform synthesis** uses digitized samples of the notes to produce a more realistic sound.

There are many CDs in multimedia devices such as CD-ROM, CD-R, and CD-ROM XA, etc.

add-on 外加的

analog ['ænəlɔg] n. 模拟

digital-to-analog converter 数模转换器

FM(Frequency Modulation) 频率调制

FM synthesis FM合成法

waveform synthesis 波形合成法

◇ **CD-ROM**

CD-ROM 只读光盘

CD-ROM is a small plastic disk that is used as a high capacity ROM storage device **that can store** 650MB of data. Data is stored in **binary** form as holes etched on the surface, which can be read by a laser. CD-ROM drives normally have an access time of between 150~300milliseconds, compared to under 15ms for a fast hard disk drive. A **single-speed** disk spins at 230rps.

binary ['bainəri] adj. 二进制的

single-speed 单速

◇ **CD-R**

CD-R 可擦写光盘

This is the technology that allows a user to write data and read data from a CD-R disk. A CD-R disk can be played in any standard CD-ROM drive but needs a special CD-R drive to write data to the disk.

◇ **CD-ROM XA**

This is the enhanced CD-ROM format developed by Philips, Sony and Microsoft that allows data to be read from the disk at the same time as audio is played back. The standard defines how audio, images and data are stored on a CD-ROM disk to allow sound and video to be **accessed** at the same time. CD-ROM XA drives can read a Kodak Photo CD disk. CD-ROM XA disks can be played on a CD-I player. A CD-ROM XA disk can be read in a standard CD-ROM player, but requires a special CD-ROM XA controller card.

enhanced [in'hɑ:nst] adj. 增强的

access ['ækses] vt. 获取，访问

◇ **Video capture board**

video capture board 视频采集卡

A video capture board is a high speed digital sampling circuit which stores a TV picture in memory so that it can then be processed by a computer.

◇ **Graphics accelerator board**

graphics accelerator board 图形加速卡

Graphics accelerator board is a specialized expansion board containing a graphics coprocessor as well as all of the other circuitry found on a video **adapter**. Sometimes it is called a **video accelerator board.** Transferring most of the graphics processing tasks from the main processor to the graphics accelerator board improves system performance considerably, particularly for Microsoft Windows users.

adapter [ə'dæptə] n. 适配器

video accelerator board 视频加速卡

You can see that multimedia is by no means just one of the preceding technologies. Because of what hardware can and cannot do, it is often a trade-off between a certain number of images, audio sound, video and text.

.End.

Key Words

animation 动画
transfer rate 传输率
synthesis 合成
coprocessor 协处理器
sample 采样
bandwidth 宽带
adapter 适配器
video capture board 视频采集卡
audio capture board 音频采集卡
graphics accelerator board图形加速卡
digital-to-analog 数/模
expansion 扩展
access time 访问时间
joystick 操纵杆
voice synthesizer 声音合成器
algorithm 算法
video accelerator board 视频加速卡
graphics accelerator board 图形加速卡
waveform synthesis 波形合成

参考译文 技能2 多媒体设备

多媒体计算机广泛应用于各个领域。它的出现是计算机领域的一次变革。多媒体计算机需要比主流计算机功能更强大，至少多媒体计算机决定了主流计算机的发展。多媒体将声音、图形、动画、视频和文本结合在一起，它使计算机在开发应用方面发挥更大的作用。多媒体计算机可以运行多媒体应用程序，通常装有声卡、CD-ROM驱动器及高分辨率的彩色显示器。此外，还包括扫描仪、数码相机等。普通计算机和多媒体计算机大不相同，主要是声卡和光盘驱动器的区别。光盘是多媒体的主要存储设备和交换媒体。

中央处理器	英特尔 i7 8700K CPU 8代六核
主板	技嘉Z370 AORUS Gaming 5
内存	金士顿DDR4 2133 16GB
硬盘	三星 850 EVO 1TB SATA3
显卡	影驰GTX 1080 GAMER 1683(1822)MHz/10000MHz 8G/256Bit
显示器	三星U28E590D 28英寸4K高清LED背光液晶
光驱	LG 24倍速 SATA接口 内置DVD
机箱	金河田峥嵘Z2
电源	冷酷至尊 550W V550
音箱	惠威 M200MKIII+ HIFI 2.0 蓝牙

◇ 声卡

声卡是一个用数模转换器或FM合成芯片，将数字数据转换成模拟声音信号的外接式设备。PC声卡有三种主要标准：AdLib、Sound Blaster及Windows兼容。多数PC声卡带有可将MIDI数据转换成声音的电子线路。MIDI数据生成声音的方法有两种：FM合成法，通过调制基本载波的频率来仿真音符；波形合成法，使用音符的数字采样来产生仿真的声音。

目前在多媒体设备中有多种光盘，如：CD-ROM、CD-R和CD-ROM XA等。

◇ 只读光盘

CD-ROM是一种小塑料盘，作为一种高容量的只读存储设备，它可以存储650MB的数据。数据以二进制格式存储，二进制数字用光盘表面烧蚀的空间表示，可以用激光器阅读。不同于15毫秒以下的快速硬盘驱动器，CD-ROM驱动器的访问时间一般在150～300毫秒之间，单速光盘的转速为230转/秒。

◇ 可擦写光盘

CD-R是一种允许用户读写的光盘。CD-R光盘可以在任何标准的CD-ROM驱动器上播放，但需要一种专用的CD-R驱动器才能将数据写入光盘中。

◇ 光盘驱动器的扩展结构

它是Philips、Sony及Microsoft公司开发的增强型CD-ROM格式，在音频回放的同时可以从光盘上读取数据。它定义了音频、图像和数据在CD-ROM光盘上的存储方式，以便允许同时访问声音和影像，CD-ROM XA驱动器可以阅读Kodak Photo CD格式的光盘。CD-ROM XA光盘可以在CD-I 播放器上播放，并可以在标准CD-ROM播放器上阅读，但需要专用的CD-ROM XA控制卡。

◇ 视频采集卡

视频采集卡是一个把电视图像存储在存储器中的高速数字采样电路，以便后期由计算机进行处理。

◇ 图形加速卡

图形加速卡是一种特殊的扩展卡，卡上有一个图形协处理器以及其他视频适配器上的电路，有时也被称为视频加速卡。图形加速卡承担了从主处理器到图形加速卡的大部分图形处理任务，从而显著改善了系统性能，特别是对微软Windows的用户而言。

多媒体绝不是一种过时的技术，我们要根据硬件的能力，权衡选择图像、声音、影像和文本。

Fast Reading One Photoshop

Photoshop has proven useful in a wide range of professional fields including science, art and design. Astronomers rely on the processing power of Photoshop to deal with the massive amounts of photographic data. Animators use Photoshop to create visually appealing, multi-layered characters and environments. Graphic and web designers rely on layers and Photoshop's other creative features to design images and products that are both attractive and functional.

Then, what is Photoshop? Adobe Photoshop is an image editing program used to perform a variety of functions in the graphics, photography and digital art worlds. And it was developed and published by Adobe Systems. It is commonly used by graphic designers because he can create a variety of image types, and add or modify or edit any kind of pictures.

Photoshop software is used to edit and modify a wide range of image types. One of its more common uses is basic photo editing. The editing tools available in Photoshop allow users to perform simple modifications like red-eye removal, lighting and color level adjustments, as well as more complex revisions like the layering of multiple photo elements to create a single image. This process involves a specific feature of Photoshop called layer.

Photoshop layers enable the user to arrange and rearrange images over top of one another. Each separate image has its own layer that can be moved back and forth across the image layer beneath it. Layers can also be used to add and remove visual effects like filters and lighting.

Photoshop also allows users to create original images like digital paintings, illustrations and other unique graphics. Digital painting is an art technique that involves recreating traditional painting styles such as watercolor, ink and oil through the use of computer software and digital tools. Photoshop offers a steadily-expanding set of virtual brushes and pens that can accurately reproduce many different physical drawing and painting techniques. Digital artwork is most commonly found in the films and video games.

How to use Photoshop to convert a photo to a canvas-like painting? Many people put photos that they like on the wall for their friends and family members to enjoy. Doing this does not require hiring a professional graphics artist. If you have access to Photoshop, you can easily create this effect on your own computer.

◇ Transfer the image you want to apply a canvas effect to onto your computer from a digital camera.

◇ Start Photoshop.

◇ Select "File" from the menu above and choose "Open."

◇ In the dialog that pops up, go to and open the photo you loaded.

◇ Select "Filter" from the menu and choose "Artistic."

◇ From the options, choose "Dry Brush."

◇ Adjust the "Size," "Detail" and "Texture" sliders until you like the result. Click on "OK."

◇ Click "Filter" in the menu and click "Texture."

◇ From the options, click "Texturizer."

◇ Adjust the "Scaling" and "Relief" sliders until you like the effect. Click on "OK."

◇ Save your work.

.End.

参考译文 PS图像处理软件

PS图像处理软件已经在许多专业领域比如科学、艺术与设计领域证明其是有用的。天文学家依靠PS图像处理软件的处理能力来处理影像资料。动画制作者使用PS图像处理软件创建富有视觉吸引力、多层次的人物和环境。平面和网络设计者依赖图层和PS图像处理软件的其他创造性特征来设计具有吸引力和多种用途的图像和产品。

那么，什么是PS图像处理软件？PS图像处理软件是一个图像编辑程序，在图形、摄影与数字艺术处理方面功能强大。由Adobe公司开发并发行。平面设计师通常会使用这一软件，因为他可以利用该软件创建各种类型的图像并添加、修改或编辑任何图片。

PS图像处理软件可以编辑和修改一系列的图像类型。它的一个比较常见的用途是基本的照片编辑。PS图像处理软件中的编辑工具允许用户做简单的修改，如去除红眼、亮度和色彩的调整，还可以进行更加复杂的修改，如将多种照片元素叠加在一起，形成新的图像。这个过程要用到一个叫做图层的PS图像处理软件的特定功能。

PS图像处理软件的图层允许用户反复将一个图层置于另一个图层之上。每个单独的图像都有自己的图层，可以自由地与其下面的图层交换位置。图层也可用于添加和删除像滤镜和亮度这样的视觉效果。

PS图像处理软件还允许用户创建像数字绘画、插图和其他独特图形的原始图。数字绘画是一种艺术技法，通过使用计算机软件和数字化工具再现传统的绘画风格，比如水彩画、水墨画和油画。PS图像处理软件提供了一个可不断扩大的虚拟毛笔和钢笔，可以准确地再现许多不同的物理绘画技巧。在电影和视频游戏中，数字艺术作品是最常见的。

如何使用PS图像处理软件把照片转换为油画？许多人把照片挂在墙上与他们的朋友和家人一同分享。这样做不需要雇佣一个专业的制图艺术家。如果你能使用PS图像处理软件，便能在自己的计算机上轻松地营造出这种效果。

◇ 从数码相机上把你想营造出油画效果的图像转移到你的电脑上。
◇ 启动PS图像处理软件。
◇ 选择“文件” 菜单中的 “打开”。
◇ 在弹出的对话框中，打开你加载的照片。
◇ 选择“滤镜” 菜单中的“艺术”。
◇ 从选项中选择“干画笔”。
◇ 调整“大小”“细节”和“纹理”滑块直到出现你喜欢的结果。单击“OK”。
◇ 单击“滤镜”菜单中的“纹理”。
◇ 从选项中，单击“纹理构成”。
◇ 调整“材质缩放”和“材质凹凸”滑块直到出现你喜欢的效果。单击“OK”。
◇ 保存你的操作。

Fast Reading Two　Video Compression

How to understand video compression? If you've ever worked with still images, and have seen the difference in storage size between a comparable BMP image, and a JPEG picture, you've started to understand the reason for video compression. By discarding redundant information, compression can save storage space needed to transmit an image. It would take enormous resources to store, process and transmit all that information in an uncompressed manner.

Compression is the conversion of data to a format that takes fewer bits. The size of data in compressed form (C) relative to the original size (O) is known as the compression ratio (R) in the formula R=C/O. When you back up your computer data files and compress them, some redundant information will be eliminate. If you can restore them to an identical state, the process is called lossless compression. If you can only generate an approximation of the original image or video, this process is called lossy compression.

The encoder refers to the hardware or software devices that compile and convert the signals (such as bit stream) or data, and it is used to communicate and storage the signals. Each encoder has a fixed light source opposite a light detector. The decoder is DAC that can convert the digital signal into analog signal. A codec can refer to the mathematical operation mechanism used to encode or decode, for example, a JPEG picture. Codec is often used in video conferencing and streaming media applications.

There are several inter frame compression techniques. As mentioned before, MPEG-2 employs both intra frame and inter frame compression. The inter frame compression is based on a Group of Pictures (GOP) which repeats every half second. The so-called GOP is a group of consecutive frames. The frame is divided into "I", "P" and "B" by the code MPEG. "I" is the internal coding frame, "P" is the forward predicted frame, "B" is a bi-directional interpolation frame. After the "I" frame, the next frame to be encoded is a P or predictive frame, which anticipates changes in the video, discarding redundant information. Between the "I" and "P" frames, "B" seems both backwards and forwards. This method creates very efficient compression and works well.

When the color video is compressed, or a film is converted to video to produce the highest quality, each of the color components is sampled in a ratio of 4:4:4 namely red, green and blue. In video there are two additional forms of compression, intra frame and inter frame, sometimes intra frame is called spatial compression, and inter frame is called temporal compression. Let's look at intra frame first, which is called the most popular intra frame compression, that is, Discrete Cosine Transform (DCT). A Discrete Cosine Transform (DCT) expresses a finite sequence of data points in terms of a sum of cosine functions oscillating at different frequencies. DCTs are important to numerous applications in science and engineering, from lossy compression of audio (e.g. MP3) and images (e.g. JPEG) to spectral methods for the numerical solution of partial differential equations.

This is an illustration of the implicit even/odd extensions of DCT input data (e.g Pic 5.1), there are the four most common types of DCT (types I-Ⅳ) when data points (red dots) N equals 11. For compression, it turns out that cosine functions are much more efficient than sine functions, whereas for the cosine functions of the differential equations express a particular choice of boundary conditions. It's a complicated process. The procedure is based on the premise that pixels in an image share a relationship with their neighboring pixels. The redundancy in information is used to compress the image. It is sampled in 8×8 pixel blocks producing a cosine that enables the processing to discard any details in the picture that are undetectable to the human eye. The process was honed through trial until there was no detectable difference between the original image and the one that was compressed. Compression works best when there is a lot of redundancy in the image.

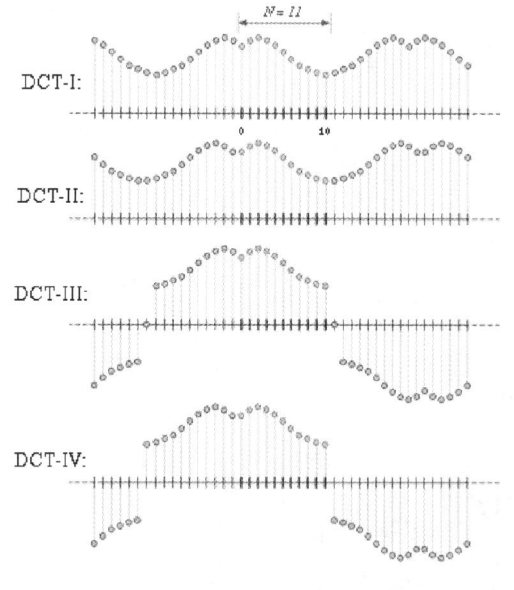

Pic 5.1

.End.

参考译文 视频压缩

如何理解视频压缩？如果你曾用过静止图像，并看到过BMP格式图像和JPEG格式图片在存储大小上的差异，你就会理解视频压缩的原因。通过丢弃冗余信息，压缩可以节省用于传输图像的存储空间。以未压缩的方式存储、处理和传输所有信息将需要巨大的资源。

压缩可将数据转换成占用较少比特的数据格式。压缩形式中数据的大小（C）与原始大小（O）的比值即为压缩比（R），公式为R=C/O。当你备份电脑数据文件并将其压缩的时候，一些冗余信息将会被删除。如果数据可以恢复到完全相同的状态，这个压缩过程就被称为无损压缩。如果你只能解压产生一个近似的原始图像或视频，这个压缩过程就被称为有损压缩。

编码器指的是将信号（如比特流）或数据进行编制、转换的硬件或软件设备，编码器用于通信和存储信号。每个编码器都有一个固定光源，与之相反的另一侧有光线侦测器。解码器就是把数字信号转化成模拟信号的数模转换器。编码解码器指的是用于编码或解码，如JPEG格式图片的数学运算结构。编码解码器经常用在视频会议和串流媒体等应用中。

现如今，有几种帧间压缩技术。如前所述，MPEG-2采用帧内和帧间压缩。帧间压缩基于每半秒重复一次的画面组（GOP），所谓的GOP就是一组连续的帧。MPEG编码将帧分为I帧、P帧、B帧。I帧是帧内编码帧，P帧是前向预测帧，B帧是双向预测内插编码帧。在I帧后，下一个将被编码的帧是P帧或预测帧，它能预测视频的变化，丢弃多余的信息。在I帧和P帧之间，B帧看起来既可以向前又可以向后。此方法能创建非常高效的压缩，效果很好。

当压缩彩色视频或当一部电影被转换为最高质量的视频的时候，每种颜色成分都按4:4:4（红:绿:蓝）的比例取样。视频文件有两种额外的压缩形式：帧内和帧间。有时帧内被称为空间压缩，帧间被称为时间压缩。让我们首先看看帧内，它被称为最受欢迎的帧内压缩，即离散余弦变换（DCT）。根据一定数量的余弦函数的振荡，离散余弦变换表示在不同频率的数据点的有限序列。离散余弦变换在科学和工程中有许多重要的应用，从音频压缩（如MP3）、图像压缩（如JPEG），一直到偏微分方程中的数值解谱方法。

这是一个内隐奇/偶DCT输入扩展数据说明(图5.1)，当数据点（红点）N等于11时，有四种最常见的离散余弦变换类型（类型I-Ⅳ）。就压缩而言，该图证明了余弦函数比正弦函数更有效，而微分方程中的余弦函数则表示一个边界条件的特定选择。这是一个复杂的过程。该程序基于图像中的像素与其相邻像素共享它们之间的关系这一前提。冗余的信息被用于压缩图像。这是8×8像素块产生余弦的一个采样，余弦使程序丢弃了许多人眼观察不到的细节图像。这个过程是通过反复的试验，直到原始图像和被压缩图像之间没有可以检测到的差异为止。冗余信息越多，图像压缩效果越好。

Ex 1 **What is Multimedia? Try to describe it.**

Ex 2 Fill in the table below by matching the corresponding Chinese or English equivalents.

algorithm	
	音频
JPEG	
	视频
MPEG	
	像素

Ex 3 Choose the best answer to the following questions according to the text we learnt.

1. In fact, multimedia is just two media: __________and___________.

   A. sound, video

   B. pictures, video

   C. animations, text

   D. sound, pictures

2. Which of following statements is incorrect?

   A. Multimedia is a kind of computer technology

   B. Multimedia is the combination of PCs and TV

   C. Multimedia is a mix of machine and ideas

   D. Multimedia is a form of communication

3. ____________ standards are the main algorithms used to compress video and have become international standards since 1993.

   A. MPEG

   B. JPEG

   C. MIDI

   D. JPEG and MPEG

4. Video applications deal with the so-called ___________ in order to define images.

   A. RGB

   B. color spaces

   C. YUV

   D. pixel

5. The simplest representation of digital video is a ___________.

   A. rectangular

   B. pixel

   C. sequence of frames

   D. grid

Part B Practical Learning

Training Target

In this part, students must finish two special tasks in English environment, under the guidance of the Specialized English teacher. First the teacher must divide the students into several groups.

Task One Configure Appropriate Pictures and Text for the Goods

In this part, under the guidance of the Specialized English teacher, students will describe the Apple iPhone 5s (Latest Model, Pic 5.2) smartphone according to the product information. Say something about the product identifiers, key features, battery, display, other features and dimensions.

Pic 5.2 Lastest Model

Product Information	
This is an iPhone 5s with high-efficiency and high-quality. The key feature of the Apple mobile phone is 4-inch multi-touch widescreen, which supports 1136×640-pixel resolution. The phone is powered by a 64-bit Apple A7 processor and an M7 motion coprocessor. The iPhone 5s is fully supported by the iOS 7, which makes you more enjoyable with its great features. Moreover, the rear iSight camera on this Apple mobile phone lets you capture beautiful photos and it supports Full HD video recording. What's more, this smartphone supports 802.11n Wi-Fi, Bluetooth 4.0, and 4G/LTE, which eases your sharing and surfing experience.	
Product Identifiers	
Brand	Apple
Carrier	AT&T
Family Line	Apple iPhone
Model	5s

UPC	885909727469
Type	Smartphone
Key Features	
Storage Capacity	16 GB
Color	Gold
Network Generation	2G, 3G, 4G
Network Technology	GSM / EDGE / CDMA EV-DO Rev. A and Rev. B / UMTS / HSPA+ / DC-HSDPA / LTE
Style	Bar
Camera	8.0 MP
Battery	
Battery Type	Rechargeable Li-Ion Battery
Battery Talk Time	Up to 600 min
Battery Standby Time	Up to 250 hr
Display	
Display Technology	Retina
Diagonal Screen Size	4 in.
Display Resolution	1136×640 pixels
Other Features	
Touch Screen	Yes
Bluetooth	Yes
Digital Camera	Yes
GPS	Yes
Email Access	Yes
Internet Browser	Yes
Speakerphone	Yes
Dimensions	
Height	4.87 in.
Depth	0.3 in.
Width	2.31 in.
Weight	3.95 oz

Task Two Beautify the Online Store Page

In this task, first, students can apply what they have learned to beautify the online store page. Secondly, students can publish the online store page after the design is finished.

Building your own website isn't easy, let alone beautifying the online store page (Pic 5.3). And luckily there are many tools to help you along the way—some free, some quite expensive. Anyway, be prepared to learn a lot about graphic design, typography, and color theory, as well as web languages like HTML and CSS.

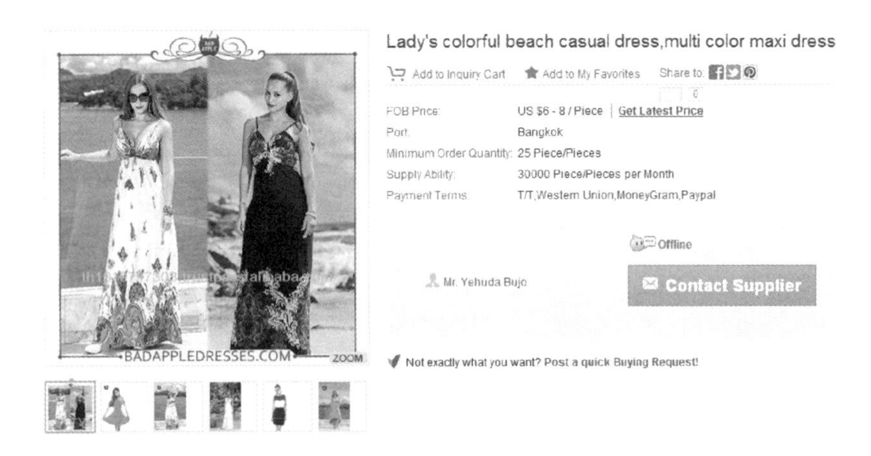

Pic 5.3

In order to attract more customers, you need to use color and design elements purposefully to beautify the online store page.

Step 1

Know what the website looks like before beautifying the online store page. The clearer an idea you have, the better you'll design.

Step 2

Sketch out layout. Typically, websites have at least two major layouts: a front page layout and the layout for all the inside pages.

Step 3

Install a high-powered graphics editing program, such as Adobe Photoshop, Corel Paint ShopPro, or Gimp. These programs are quite complicated, so be prepared to take some time to learn them well.

Step 4

Select a color scheme for your website. This step can actually be quite fun because you can find several free color scheme design tools online.

Step 5

Choose a typeface for your site. Text is an integral part of your site's design, and choosing a smart font can contribute to beautifying the online store page—but keep in mind that a pretty site without a readable text is useless. Only a few fonts are "web-safe", you can adjust the typeface's size, line spacing, and word gap to distinguish it from others. Most websites use the so-called "web-safe" fonts to make sure that all users "see" the same style when visiting a page.

Step 6

Design the site by using your graphics program. Your goal here is to create an entire visual representation of a webpage by using the layout, color scheme, and fonts you've chosen, and you can get attractive backgrounds by using subtle gradients.

Step 7

Convert your finished design to HTML and CSS markup. To do this, you need to learn two kinds of languages which have numerous free tutorials online. For example, a sidebar.css file might look like Pic 5.4:

```
CSS
/*
<aside>
    <div class="widget">
        <h5 class="widget-title">
        <p></p>
    </div>
    <!-- other widgets -->
</aside>
*/

aside { background-color: #ccc; }
aside .widget { background-color: white; padding: 10px; }
aside .widget h5 { border-bottom: 1px solid black; }
/* etc. */
```

Pic 5.4 A sidebar.css file

The markup is generated by JavaScript, and the CSS is specific to a plug-in or a third-party plug-in.

Step 8

Publish the site once the design is finished, then enjoy the following compliments! File synchronization is easy, just sync or copy folders over network. Build 3D photos by taking advantage of digital photos' robustness, accuracy and efficiency.

Exercise

Ex 1 **According to the following picture, first of all configure the proper text for the GATINEAU. Secondly, fill in the form with the given information of the GATINEAU.**

GATINEAU
DefiLift 3D Perfect Desian Performance
Volume Cream 50ml

GATINEAU DefiLift 3D Perfect Design Performance Volume Cream 50ml

Brand:	________________
Volume:	________________
Directions:	积姬仙奴Defilift 50ml立体丰盈霜适用于脸部肌肤的早、晚呵护，使用时应特别注意太阳穴、脸颊和下巴等部位。 GATINEAU DefiLift 3D Perfect Design 50ml ________________ to the face paying special attention to the ________________.
Ingredients:	1. Re-plumping the cells with moisture, improving the appearance of the facial contours. 翻译：________________。 2. Containing protein extracted from wheat, corn and soya. 翻译：________________。 3. Exclusive plant protein complex forms a three-dimensional micro-network on the skin. 翻译：________________。

Ex 2 **If you want to sell your products on a free web-store where people can choose products and buy them by transferring money to your account. You need to build your online store first. Please describe how to build a free online store in your own words.**

Part C Occupation English

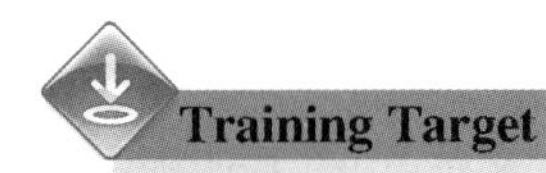

In this part, students are supposed to practice the dialogue, which may occur occasionally in their routine. Communication, technical support, and useful instruction may bring you extra benefit.

How to Set up Webpage 如何建立网页
Post: Webpage Designer （岗位：网页设计师）

A: Hello. I want to publish a story I've written, and create an electronic magazine related to one of my hobbies. Could you tell me how to create webpage?
我想发布一个我写的故事，并且想创建一个与我的爱好相关的电子杂志。您能告诉我如何创建网页吗？

B: Of course. Once you get the hang of it, you'll find it interesting. I'll tell you when you turn on your computer.
当然可以，一旦你知道它的窍门，就会觉得它很有趣。你打开电脑我再告诉你。

A: Ah! I can't wait.
啊！我等不及了。

B: First you need to create a text document containing the words that you want to appear on your webpage. You can save the document as "Text Only" as long as you give your document a name ending in .html or .htm. Then type in your story or whatever you want to publish.
首先你需要创建一个包含你想呈现在网页上的文字的文本文件。让文档以.html 或.htm结尾，将文档保存为“纯文本”。然后键入你的故事或你想发布的任何东西。

A: Ok, that's fine. Also, how can I organize my "Text Only"?
好的。那我如何组织我的“纯文本”呢？

B: Here are some tips for organizing your text: Begin with a title and an attention-grabbing greeting or introduction, so people will know immediately what your page is about and will want to read more.
这里有一些组织文本的技巧。以一个标题或一个引人注目的问候或介绍开始，这样人们就会立刻知道你的网页是什么内容，会继续读更多的东西。

A: Do I need to learn basic HTML tags to create my webpage?
我需要学习基本的HTML标签来创建我的网页吗？

B: Yes, you do. Now you need to insert some formatting tags in your document, which will tell an Internet browser how to arrange your words and pictures on the screen. These tags are HyperText Markup Language (HTML).
是的，你需要。现在你需要向你的文档中插入一些格式标签，这些格式标签将告诉浏览器在屏幕上如何安排你的文字和图片。这些标签被称为超文本标记语言（HTML）。

A: How to insert formatting tags in my document?

如何在我的文档中插入格式标签？

B: Remember to start your document with <html> and end it with </html>. The “title” that you type between the tags <title>...</title> will appear at the very top of the browser window when your page is viewed.

记住要用<html>开始，用</html>结束你的文件。写在<title>和</title>之间的标题在浏览网页时会出现在浏览器窗口的顶部。

A: How does HTML become a webpage?

HTML是如何转换成一个网页的？

B: You may feel somewhat baffled after reading that long list of formatting tags. In fact, it’s very easy to look at a real HTML document. You can view the “Source Document” behind any webpage by going to the “View” menu and selecting “View Source”. When you do this, you’ll see the document with HTML tags that create the page currently in the window of your browser.

阅读完长长的格式标签清单后，你可能会感到有些困惑，事实上，查看HTML文档是很容易的。您可以通过进入“视图”菜单、选择“查看源代码”的方式查看任何网页的“源文件”。这样，你就能看到创建了当前网页的HTML格式标签了。

A: Could you tell me how to open a webpage?

您能告诉我如何打开网页吗？

B: Yes. Under the “File” menu, select “Open File”. From here you will find your document’s name and then click on “Open”. Your document will appear in your window.

好的。在“文件”菜单下，选择“打开文件”。之后你就能够找到你的文件名，然后点击“打开”。你的文件就会出现在你的窗口了。

A: It looks like a real webpage!

看起来像真正的网页了！

B: Yes, it is. But you need to publish your webpage on the Internet. If you don’t publish it, the viewers won’t be able to visit it unless they know your page —and the exact URL!

是的。但是你需要在互联网上发布你的网页。如果你不发布，除非浏览者知道你的网页，知道确切的URL，否则他们是不能访问的。

A: How to publish my webpage? And how will they find it?

如何公开我的网页，别人怎样找到它呢？

B: Your school may have an account with an Internet provider, which allows you to put your webpage on a server so that other people can read it. If not, you can register your site with a search engine such as Google or Yahoo. Once you add your URL to one of these databases, people who are interested in the topic of your page will get a listing of your site when they are surfing the Internet. You have set up a real webpage.

你的学校可能会有一个互联网服务提供商的账户,能允许你把你的网页放到服务器上，让其他人看到。如果没有，你可以在谷歌或雅虎等搜索引擎上注册你的网站。一旦你把URL添加到一个数据库，人们进行搜索的时候，对你的页面主题感兴趣的人就会得到你的站点列表。这样你就创建了一个真正的网页。

A: Oh, That's perfect. I see. Thanks very much.
啊，真是太好了。我明白了。非常感谢。
B: You're welcome.
不客气。

Word Building

前缀/后缀由一个或几个字母组成，放在词根或单词之前/之后，组成一个新词。

(1)non-（前缀）：不，无
smoker:抽烟的人 ———— nonsmoker: 不抽烟的人
effective:有效的 ———— noneffective: 无效的

(2)in- （前缀）：不、无、非
sensible: 有感觉的 ———— insensible: 无知觉的
correct:正确的 ———— incorrect: 错误的

(3)-fy (后缀)：表示“使……化”
beauty:优美 ———— beautify:美化
pure:纯净的 ———— purify:净化

(4)-en(后缀)： 表示“使……”
quick:迅速的 ———— quicken:加快
weak:虚弱的 ———— weaken:变弱

(5)-th(后缀)：动作，性质，过程，状态
deep:深的 ———— depth:深度
wide:宽的 ———— width:宽度

Ex Translate the following words and try your best to guess the meaning of the word on the right according to the clues given on the left.

capable	有能力的(形容词)	incapable ________
complete	全部的(形容词)	incomplete ________
sexual	性别的(形容词)	nonsexual ________
existent	存在的(形容词)	nonexistent ________
significant	重要的(形容词)	signify ________
simple	简单的(形容词)	simplify ________
soft	软的(形容词)	soften ________
hard	坚硬的(形容词)	harden ________
true	真实的(形容词)	truth ________
long	长的(形容词)	length ________

Ex 1 What's a multimedia computer? What are the differences between MPC and PC?

Ex 2 Fill in the table below by matching the corresponding Chinese or English equivalents.

animation	
	数模转换器
single-speed	
	图形加速卡
video capture board	
	FM合成器

Ex3 Choose the best answer to the following questions according to the text we learnt.

1. If you want tell an ordinary computer from a multimedia computer, the only things are__________.

   A. soundboard and CD-R

   B. voice synthesizer and video capture board

   C. soundboard and CD-ROM driver

   D. voice synthesizer and graphics accelerator board

2. According to the text, multimedia PC can run multimedia applications, the following statements are true except ____________.

   A. CD-ROM drive　　B. network card

   C. sound card　　D. color monitor

3. According to the text, you can find the following major standards for PC sound card except ______.

   A. AdLib　　B. Windows-compatible

   C. MIDI　　D. Sound Blaster

4. There are varies technologies or algorithms used to create sounds in music synthesizers. Two widely used techniques in PC sound card are ________.

   A. Frequency Modulation (FM) synthesis and Digital synthesis

   B. Waveform synthesis and Digital synthesis

   C. Waveform synthesis and Sound synthesis

   D. Frequency Modulation (FM) synthesis and Waveform synthesis

5. __________ is a high speed digital sampling circuit which stores a TV picture in memory so that it can then be processed by a computer.

   A. Graphics accelerator board　　B. Sound card

   C. Video accelerator board　　D. Video capture board

Project Six

Creating Database for the Goods

Part A Theoretical Learning

Part B Practical Learning

Part C Occupation English

Part A Theoretical Learning

Training Target

In this part, our target is to improve the speed of reading professional articles and the comprehension ability of the reader. We have marked specialized vocabulary key words in some paragraphs so that the reader can quickly grasp the main idea of the sentences and paragraphs.

Skill One Foundation of Database System

1. Database system application

Database Systems are now used in various aspects of society, such as **government apparatus**, universities, airlines, banking, telecommunication and manufacturing. In government apparatus, database can be used to know information resources of human affairs and carders in various aspects as well. In banking, database can be used for customer information, **accounts**, **loans**, and banking **transactions**. In airlines, database can be used for **reservations** and schedule information. In universities, database can be used for student information, course **registrations** and grades. **In the management of economy, database can process statistic data, analyze and obtain a result so as to guide the enterprises to develop rapidly.** Previously, very few people interacted directly with the database system. Till the late 1990s, the visit to database for user increased with the development of the Internet work. For example, when you access a network station and **look through** the content on it, in fact, you are accessing the data stored in database. When you access an on-line shopping website, information about your order for goods may be **retrieved** from a database. Database plays an import role in most enterprises today.

database ['deitəbeis] n. 数据库
government apparatus 政府机构（关）
account [ə'kaunt] n. 账户
loan [ləun] n. 贷款
transaction [træn'zækʃən] n. 交易
reservation [rezə:veiʃən] n. 订票
registration [redʒi'streiʃən] n. 注册
statistic [stə'tistik] adj. 统计的
look through 浏览
retrieve [ri'tri:v] vt. 检索，恢复

2. Basic database conceptions

(1) Data: Data is a collection of facts made up of numbers, characters and symbols, stored on a computer in such a way that the computer can process it. Data is different from information in that they are formed of facts stored in machine-readable form. When the facts are processed by the computer into a form, which can be understood

by people, the data becomes information.

(2) Database: Database is a collection of related objects, including tables, forms, reports, **queries**, and **scripts**, created and organized by a database management system (DBMS). A database can contain information of almost any type, such as a list of magazine **subscribers**, personal data on the space shuttle astronauts, or a collection of graphical images and video clips.

query ['kwiəri] n. 查询

script [skript] n. 脚本

subscriber [səb'skraibə] n. 订购者

(3) Database management system: Database management system is a software that controls the data in a database, including overall organization, storage, retrieval, security, and data **integrity**. **A DBMS can also format reports for printed output and can import and export data from other application using standard file formats.** A **data-manipulation** language is usually provided to support database queries.

integrity [in'tegriti] n. 完整性

data-manipulation 数据控制

(4) Database model: **Database model is the method used by a database management system (DBMS) to organize the structure of the database.** The most common database model is the relational database.

(5) Database server: Database server is any database application that follows the client/server architecture model, which divides the application into two parts: a front-end running on the user's workstation and back-end running on a server or host computer. The front-end interacts with the user, collects and displays the data. The back-end performs all the **computer-intensive** tasks, including data analysis, storage, and manipulation.

computer-intensive 信息密集的

(6) Relational database: Relational database is a database model in which the data always appears from the point of view of the user to be a set of **two-dimensional** tables, with the data presented in rows and columns. The rows in a table represent records, which are collections of information about a specific topic, such as the entries in a doctor's patient list. The columns represent fields, which are the items that make up a record, such as the name, address, city, state, and zip code in an address list database.

two-dimensional adj. 二维的

.End.

database n.数据库
model n.模型
enterprise n.企业
online n.在线
object n.对象
symbol n.符号
query n.查询
two-dimensional adj.二维的
DBMS (Database Management System) 数据库管理系统
relational database 关系数据库

参考译文 技能1 数据库基础

1.数据库系统的应用

当前，数据库被广泛应用到社会的各个领域，如政府机关、学校、航空、银行、电信、制造业等各行各业。在政府机关，使用数据库可以掌握各种人事、干部信息资源；在银行，数据库可用于存储客户的资料、账户、贷款以及银行的交易记录；在航空业，数据库可用于存储订票和航班的信息；在学校，数据库可用于存储学生的个人资料、课程注册和成绩记录等；在经济管理上，数据库可以进行数据统计、分析，得出结果以便指导企业更好地发展。以前，很少有人直接和数据库打交道。到20世纪90年代末，互联网的发展增加了用户对数据库的访问。比如，当你访问一个网站，浏览上面的内容时，实际上你是在浏览存储在某个数据库中的数据。当你访问一个在线购物网站时，你所订购的商品的信息就会从某个数据库中被检索出来。今天，数据库在很多企业中扮演着非常重要的角色。

2.数据库的基本概念

(1)数据：数据是由数字、字符和符号组成的信息集合，以计算机能够处理的方式存储在计算机上。数据与信息不同，数据是以计算机可识别的方式存储的信息。当计算机把这些数据转换成人可以理解的形式时，数据就变成了信息。

(2)数据库：数据库是由数据库管理系统(DBMS)创建和组织的一组相关对象，如表、表格、报表、查询和脚本。数据库所包含的信息可以是任何形式的，如一组杂志订户名单、一份宇宙飞船上宇航员的个人资料、一组图像或视频素材。

(3)数据库管理系统：它是数据库中控制数据的应用软件，控制数据的整体组织结构、存储方式、查询、安全管理以及数据的完整性。DBMS可以编排报表打印格式，并能从其他使用标准文件格式的应用中输入、输出数据资料。数据控制语言能提供对数据库的查询支持。

(4)数据库模型：数据库管理系统组织数据库结构的方法被称为数据库模型。最常见的数据库模型是关系数据库。

(5)数据库服务器：任何遵循客户/服务器结构模型的数据库应用程序都分为两部分：在工作站上运行的前台部分和在服务器或主机上运行的后台部分。前台程序负责与用户交互，接收并显示数据。后台程序执行所有的信息密集型任务，包括数据的分析、存储和数据处理。

(6)关系数据库：它是一种数据库模型，从用户的角度来看，其中的数据都以二维表的形式出现，数据排成若干行和列。表中的行表示记录，即有关某一主题的信息集，比如医生手中的患者列表。列表示字段，是组成记录的各条信息，如地址列表数据库中的姓名、地址、城市、州和邮政编码等信息。

Skill Two An Introduction to SQL

Several different languages can be used with relational databases, the most commonly used one is **SQL**. SQL is an abbreviation of Structured Query Language, and pronounced either see-kwell or as separate letters. SQL is a standardized query language for requesting information from a database. SQL's simple commands make it easy for you to design and manage the information in your database. For example, SQL commands are classified as **DDL**, **DML** and **DCL** language commands. Now let's talk about data manipulation language (DML) commands, which you can use to insert, update, delete, and retrieve information from the tables in a database. There are two keywords: "SELECT" and "FROM" in SQL. SQL Server use DDL, DML and DCL sentences to control its program.

The original version called SEQUEL (Structured English Query Language) was designed by an IBM research center in 1974 and 1975. SQL was first introduced as a commercial database system in 1979 by Oracle Corporation. **The American National Standards Institute (ANSI) and the International Standards Organization (ISO) define software standards, including standards for the SQL language.** SQL Server 2000 supports the Entry Level of SQL-92, the SQL standard published by ANSI and ISO in 1992. The dialect of SQL supported by Microsoft SQL Server is called Transact-SQL (T-SQL). T-SQL is the primary language used by Microsoft SQL Server applications. Historically, SQL has been the favorite query language for database management systems running on minicomputers and mainframes. Increasingly, however, SQL is being supported by PC database systems because it supports **distributed databases** (databases that are spread out over several computer systems). This enables several users on a local-area network to access the same database **simultaneously**.

SQL
(Structured Query Language)
abbr. 结构化查询语言

DDL
(Data Definition Language)
数据定义语言
DML
(Data Manipulation Language)
数据操作语言
DCL (Data Control Language)
数据控制语言

distributed database
分布式数据库
simultaneously [ˌsiml'teiniəsli]
adv. 同时地

Although there are different dialects of SQL, it is nevertheless the closest thing to a standard query language that currently exists. In 1986, ANSI approved a rudimentary version of SQL as the official standard, but most versions of SQL since then have included many extensions to the ANSI standard. In 1991, ANSI updated the standard. The new standard is known as SAG SQL. SQL contains about 60 commands and is used to create, modify, query, and access data organized in tables. It can be used either as an **interactive interface** or as **embedded** commands in an application:

(1)**Dynamic** SQL statements are interactive, and they can be changed as needed. If you normally access SQL from a command-line environment, you are using dynamic SQL, which is slower than static SQL but much more flexible.

(2)**Static** SQL statements are coded into application programs, and as a result, they do not change. These statements are usually processed by a **precompiler** before being bound into the application. SQL is such a popular standard today, every major client/server application supports it. Many databases implement SQL queries behind the scenes, enabling communication with database servers in systems with client/server architecture.

interactive interface 交互界面

embed [im'bed] v.嵌入

dynamic [dai'næmik] adj. 动态的

static ['stætik] adj. 静态的

precompiler [ˌpri:kəm'pailə] n. 预编译程序

.End.

Key Words

SQL 结构化查询语言

ANSI 美国国家标准协会

ISO 国际标准化组织

DDL 数据定义语言

DCL 数据控制语言

DML 数据操作语言

data-manipulation 数据操作

client/server 客户/服务器

distributed database 分布式数据库

interactive interface 交互界面

integrity n.完整性

retrieve v.检索，恢复

simultaneously adv.同时地

embed v.嵌入

dynamic adj.动态的

static adj.静态的

precompiler n.预编译程序

参考译文 技能2 结构化查询语言简介

处理关系数据库的语言有很多种，其中最常用的是SQL。SQL是结构化查询语言的缩写，既可以读成See-Kwell又可以读成字母SQL。SQL是从数据库中获取所需信息的标准化查询语言。SQL的简单命令可使你轻松设计和管理数据库中的信息。例如，SQL命令可被分类为DDL、DML和DCL语言命令。现在我们讨论数据操作语言（DML）命令，你可以使用它在数据库的表中插入、刷新、删除和检索信息。在SQL中有两个关键字：“SELECT”和“FROM”。SQL Server使用DDL、DML和DCL来控制其程序。

早期的版本是IBM研究中心在1974年至1975年间设计的，称为SEQUEL(结构化英语查询语言)。1979年，Oracle公司第一次将SQL用于商业数据库系统。美国国家标准协会（ANSI）和国际标准化组织（ISO）定义了软件标准，其中包括SQL语言的标准。SQL Server 2000支持入门级(Entry Level)SQL-92，即由ANSI和ISO在1992年公布的SQL标准。微软SQL Server支持的SQL方言称为Transact-SQL(T-SQL)。T-SQL是微软SQL Server应用程序使用的主要语言。一直以来，SQL都是运行在小型机和大型机数据库管理系统中的最受欢迎的查询语言。PC机数据库系统逐渐也开始支持SQL，因为它支持分布式数据库（分布在几台计算机上的数据库）。这能使本地网上的几个用户同时访问同一数据库。

尽管SQL有不同的方言，它仍然是目前存在的查询语言中最接近标准查询语言的。在1986年，ANSI批准SQL的基本版本为正式标准。但之后的很多SQL版本都扩展了ANSI标准，所以，在1991年ANSI更新了基本标准。新的标准是SAG SQL。SQL大约包含60条命令，用于生成、修改、查询及访问以表的形式组织的数据。它既可以作为交互界面，又可以作为应用程序的嵌入式命令来使用。

（1）动态SQL语句是交互式的，可以在需要时进行更改。以正常方式通过命令行环境访问SQL，实际上就是在使用动态SQL，速度比静态SQL慢，但更具有灵活性。

（2）静态SQL语句以代码形式存在于应用程序内部，因此无法更改。这些语句在嵌入应用程序前通常由预编译程序进行处理。SQL是当今最流行的标准，每个大型的客户/服务器应用程序都支持它。许多数据库在后台实现SQL查询，与客户/服务器结构系统中的数据库服务器进行通信。

Fast Reading One Object-Relational Database

A database management system (DBMS) is a software program that is used to create databases. The program also allows users to update tables and records within the database.

What is an object-relational database (ORD) or object-relational database management system (ORDBMS)? An ORD or ORDBMS is a relational database management system, which allows developers to integrate the database with their own custom data types and methods.

It allows the user to define new custom data types, new functions, and operators that manipulate these new custom types. That is, ORD combines the features of both relational databases and object-oriented programming. This means that when developing these databases, you can increase your ability to sort through and locate files within these databases, you can filter them better by assigning these data types to your files, you can also retrieve files that share the same characteristics.

However, in an object-oriented database, information is represented in the form of objects as used in object-oriented programming. Some object-oriented databases are designed to work well with object-oriented programming languages such as Python, Java, C#, Visual Basic .NET, C++ and Smalltalk. Others have their own programming languages. The term object-relational database is sometimes used to describe external software products. An object-relational DBMS allows software developers to apply their own types and methods into the DBMS.

There are many advantages of object-relational database.

◇ **Extensibility**

Object-relational database's capabilities are extended with the addition of new data types, access methods and functions found in object-oriented programming. You can add string characters with alpha-numeric data types. The complex data types can combine characteristics of data types that already exist in your database. You can specify data types by the text you wish to contain or by the amount of bytes used to create it. User-defined data types can be opaque or distinctive. You can also add user-defined virtual processors.

◇ **Inheritance**

Unlike relational databases, object-relational databases allow the use of inheritance. Within inheritance, you can develop classes for your data types. These objects can inherit certain capabilities from the classes to be used in other functions of the database. These inherited properties could be something simple. Through inheritance, your data types will inherit these features of other data types.

◇ **Polymorphism**

Polymorphism in object-relational databases involves allowing one operator to have different meanings within the same database. You can connect your tables within your database by building relationships. This includes records that may all contain the same name but different information.

◇ **Encapsulation**

You would use encapsulation with object-relational databases in the form of tables. For instance, you want Table one to include name, address, telephone number and email address for your contacts. Through encapsulating the "contacts" class, you combine all of this information into this one table. So that when you query the database for this information, you generate a report in the style of a form to include all of this information.

◇ **Database Management Systems**

Object-relational databases can be used to build database management systems. You can connect them to company websites that allow your inventory records to update. The object components used on your website can make these updates when the user places an order by clicking one of the control buttons programmed to connect to these database management systems.

.End.

参考译文 对象关系数据库

数据库管理系统是用于创建数据库的软件程序。该软件程序还允许用户更新数据库中的表和记录。

什么是对象关系数据库（ORD）或对象关系型数据库管理系统（ORDBMS）？对象关系数据库或对象关系型数据库管理系统是一个关系数据库管理系统，它允许开发者用自己的自定义数据类型和方法整合数据库。

它允许用户定义新的自定义数据类型、新的功能，和使用这些新自定义类型的运营商。也就是说，对象关系数据库结合了关系数据库和面向对象编程的特点。这意味着，当你开发这些数据库的时候，你能提高分类和查找这些数据库中的文件的能力。通过向你的文件中分配这些数据类型，你能更好地过滤通过的数据类型。你还可以检索共享相同特征的文件。

然而，在一个面向对象的数据库中，信息是以对象的形式被表示在面向对象的设计程序中的。为产生良好的效果，一些面向对象的数据库用面向对象的编程语言，如Python、Java、C#、Visual Basic.NET、C++和Smalltalk来设计。另一些面向对象的数据库也有它们自己的编程语言。对象关系数据库这一术语有时也用来描述外部软件产品。对象关系型数据库管理系统允许软件开发商将自己的自定义数据类型和方法应用到数据库管理系统中。

对象关系数据库有很多优势。

◇ 可扩展性

面向对象编程中新加的数据类型、访问方法和功能，带动了对象关系数据库功能的发展。你可以用字母-数字数据类型添加字符串中的字符。在你的数据库中，复杂的数据类型能够整合已存的数据类型的特点。你可以用你想包含的文本或用来创建文本的字节数量来指定数据类型。用户定义的数据类型可以是不透明的或独特的。你还可以添加自定义的虚拟处理器。

◇ 继承

不同于关系数据库，对象关系数据库允许继承使用。在继承中，你可以开发你的数据类型的类。这些对象能从可用于数据库的其他功能的类中继承某些能力。这些继承的属性可以是简单的东西。通过继承，你的数据类型将继承其他数据类型的某些特点。

◇ 多态性

对象关系数据库中的多态性允许一个运营商在同一数据库中有不同的含义。你可以在你的数据库中通过建立关系连接你的表。这个表可能会包括全部的名称相同而信息不同的记录。

◇ 封装

在表格形式中，你将使用带有对象关系数据库的封装。例如，你想要表1中含有姓名、地址、联系人电话号码和电子邮件地址。通过封装“联系”类，你就可以把这些信息都集合到一个表中。所以当你查询数据库信息的时候，就会得到一个包含了全部信息的表格文件。

◇ 数据库管理系统

对象关系数据库可以用来建立数据库管理系统。可以将数据库管理系统连接到允许更新项目库存记录的公司网站上。当用户点击一个连接到这些数据库管理系统程序的控制按键订货时，在你网站上所使用的对象的组件就能更新这些数据库管理系统了。

Fast Reading Two Network Database

Network database refers to a database that runs in a network. It implies that the database management system (DBMS) was designed with a client or server architecture. A network database consists of a collection of records connected to one another through links. A record is similar to an entity in the E-R model in many respects. Each record is a collection of attributes, each of which contains only one data value. A link is an association between precisely two records. A network database holds addresses of other users in the network. In the network database, a single data element can point to multiple data elements and can itself be pointed to by other data elements.

This network model (Pic 6.1) has an ability to handle more relationship types. Namely, it can handle the one-to-many and many-to-many relationships. In the network model, there is always the "first" in the table, and the "last" in the table. If you need to restructure the database and add indexes, the users can use the new logic to improve the application logic. If you add a field, you have to restructure the entire table. The network database model implements application logic and limits the flexibility in the data structure.

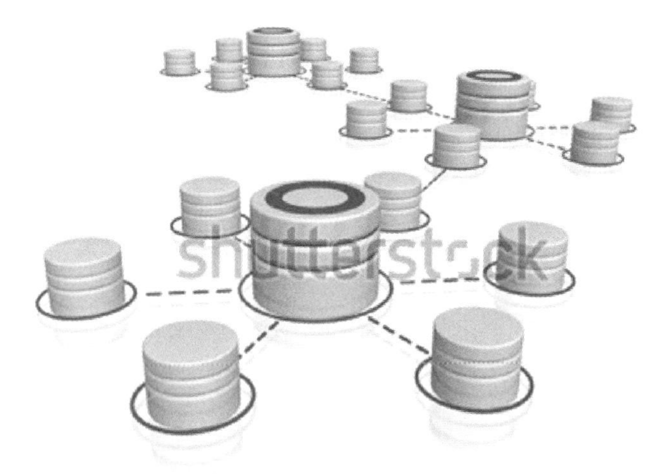

Pic 6.1 Network model

Network database is a kind of database management system in which each record type can have multiple owners, e.g. orders are owned by both customers and products. This contrasts with a hierarchical database or relational database, the former is one owner, and the latter is no explicit owner. Network database can access your work from any computer on that network. But you can't access your work if the server goes down.

◇ **Ease of data access**

In the network database, a relationship is a set. Each set comprises of two types of records. They are owner records and member records. An application in a network model can access an owner record and all the member records within a set.

◇ **Data integrity**

In the network database, no member can exist without an owner. Therefore, to ensure the integrity, a user must first define the owner record and then the member record.

◇ **Data independence**

In the network database, the data of the application programs work independently. Any changes in the data characteristics made do not affect the application program.

◇ **System complexity**

In the network database, data is accessed one record at a time. This makes it essential for the database designers, administrators, and programmers to be familiar with the internal data structures to gain access to the data.

◇ **Lack of structural independence**

In the network database, making structural modifications to the database is very difficult. Any changes made to the database structure require the application programs to be modified before they can access data. Though the network model achieves data independence, it still fails to achieve structural independence.

.End.

参考译文 网络数据库

网络数据库是一个在网上运行的数据库。这意味着网络数据库系统是用客户端或服务器架构来设计的。网络数据库是由相互链接的记录集组成的。记录在许多方面类似于一个实体的E-R模型。每个记录都是一个属性集合，每个属性只包含一个数值。一个链接正是两个记录间的关联。网络数据库包含网上其他用户的地址。在网络数据库中，一个数据元素可以指向多个数据元素，而且其本身也可以为其他数据元素所指向。

这个网络模型(图6. 1)有处理更多关系类型的能力。也就是说，它可以处理一对多、多对多的关系。在网络模型中，始终存在着“第一”表和“最后”表。如果你需要调整数据库和添加索引，用户就可以使用新逻辑改进应用逻辑。如果你添加一个字段，你必须调整整个表格。网络数据库模型实现了数据结构中的应用逻辑，并限制了数据结构中的灵活性。

网络数据库是一种数据库管理系统，数据库管理系统中的每个记录类型都可以有多个物主，如订单的物主是客户和产品。网络数据库与分层数据库或关系数据库截然不同，前者有一个物主，后者没有明确的物主。网络数据库可以通过同一网络上的任何一台电脑来访问你的工作。但是，如果服务器瘫痪了，就不能访问你的工作了。

◇ 便捷的数据访问

在网络数据库中，关系是一个集。每个集包括两种类型的记录：物主记录和成员记录。网络模型中的应用程序可以访问集内的一个物主记录和所有的成员记录。

◇ 数据的完整性

在网络数据库中，没有物主记录就没有成员记录。因此，为了确保完整性，用户必须首先定义物主记录，然后再定义成员记录。

◇ 数据的独立性

在网络数据库中，应用程序中的数据独立工作。任何数据特点的变化对应用程序都不会有任何影响。

◇ 系统的复杂性

在网络数据库中，数据的访问是一个记录一次。因此为获取数据，很重要的一点是数据库设计师、管理员和程序员们要熟悉数据的内部结构。

◇ 缺乏结构独立性

在网络数据库中，对数据库结构进行修改是很难的。对数据库结构所做的任何改动都需要在访问数据前修改应用程序。虽然网络已实现数据的独立性，但它还未能实现结构的独立性。

Exercise

Ex 1 Can you tell me in which aspects the database system is mainly used?

Ex 2 Fill in the table below by matching the corresponding Chinese or English equivalents.

DBMS	
	关系数据库
two-dimensional	
	模型
symbol	
	对象

Ex 3 Choose the best answer to the following questions according to the text we learnt.

1. Data is a collection of facts made up of numbers, symbols and __________.
   A. pictures
   B. characters
   C. words
   D. text
2. A database is a collection of related objects, including tables, forms, reports, queries, and scripts, created and organized by __________.
   A. SQL
   B. DBMS
   C. RAID
   D. DLT
3. A data management system is a __________ that controls the data in a database.
   A. hardware
   B. software
   C. OS
   D. NOS

4. What is used by a DBMS to organize the structure of the database? __________

A. A database server.

B. A database model.

C. SQL.

D. A database management system.

5. Which is the most commonly used database model? __________

A. Network database.

B. Distributed database.

C. Relational database.

D. File database.

Part B Practical Learning

Training Target

In this part, students must finish two special tasks in English environment. Under the guidance of the specialized English teacher, the students can create data information and database for their goods. First the teacher must divide the students into several groups.

Task One Create Data Information for the Goods

The first task is to grasp the basic steps and methods of DB design according to the requirements. In this task, students should know that the purpose of creating database is to provide data information for the goods.

The processes and methods of DB design:

◇ **Gather and organize the required information**

Gather all types of information to be recorded in the database, such as product name and order number.

◇ **Divide the information into tables**

Divide the information into major entities or objects, such as Products or Orders. Then each object becomes a table.

◇ **Turn information items into columns**

Decide what types of information needs to be stored in one table. Then, each item becomes a field, and is displayed as a column in the table. For example, a product table might include fields such as Name and Date.

◇ **Specify key words**

Choose each table's key words. The key words form a column, or a set of columns, which are used to specify each row. An example might be Product ID or Order ID.

◇ **Set up table relationships**

Look at each table and decide how data in one table is related to data in other tables. Add fields to tables or create new tables to clarify the relationships.

◇ **Refine the design**

Analyze design errors. Create tables and add a few sample records. Make sure the resalts are right. If not, modify your design.

◇ **Use standard rules**

Use standard rules to examine your tables, if they're not right, you need to modify them.

Task Two Use SQL to Create Database for the Goods

In this task, first of all, students can operate Microsoft SQL Server to create database. Secondly, the students can use some of the SQL tools. Thirdly, they can use SQL to design query view.

◆ **Operate Microsoft SQL Server to create database**

Step 1. Open Microsoft SQL Server Management Studio. Depending on how you installed SQL Server, if you have an icon on your desktop, click it. Otherwise, click "Start", choose "All Programs", then select "Microsoft SQL Server".

Step 2. Click the plus sign next to your server name. Then right-click "Databases".

Step 3. Click "New Database", open a New Database options screen.

Step 4. Input a name for your database in the "Database name" text box. To make programming easier, you can use one of the following name formats: a single-word name such as Database, a multiple-word name without space such as NewDatabase or a multiple-word name with underscore such as My_New_Database.

Step 5. Click "OK" to create a database. You can change the database's properties at any time by right-clicking the name of the database and choosing "Properties". All objects you create for your new database will be listed out when you click the plus sign beside your database name.

◆ **Use some of the SQL tools**

Now we can use SQL to create database for the goods.

There are a few examples, let's look at some of the "tools" that SQL provides for building queries. You can use this sort of summary of SQL language tools as a reference. You first need to decide what you want, then use these tools to produce your output. This is usually not an easy task. The design view can help you get what needs to be done, but you may also find it necessary to use the SQL view to finally get what you need.

Here are some examples.

Sample one:

```
SELECT *                    ← Selecting ALL columns;  * =  Wild card
FROM customer_t
Aggregate Functions: COUNT, MIN, MAX, SUM and AVG

SELECT COUNT  (*)           ← How many items were ordered on order# 1004?
 FROM order_line_t
     WHERE order_id = 1004 ;
SELECT COUNT (*)  AS [NumberOfItems]    ← How many items are included in the table?
FROM product_t ;

Calculate total $ value of inventory on hand:
SELECT  SUM([unit_price]*[on_hand]) AS [Total Inventory Value]
FROM product_t;
```

Sample two:

```
SELECT product_name, product_finish, unit_price
FROM product_t
WHERE product_name LIKE '%Desk'          ← In Access use * instead of %
        OR product_name LIKE '%Table'
        AND unit_price > 300 ;
VS:

SELECT product_name, product_finish, unit_price
FROM product_t
    WHERE ( product_name LIKE '%Desk'
        OR product_name LIKE '%Table' )  ← The ( ) changes the order of comparison
        AND  unit_price > 300 ;
```

```
  ...........
    WHERE unit_price > 199 AND unit_price < 301      ← Range comparison
Same as:
    WHERE  unit_price  BETWEEN  200   AND  300;      ← Range comparison
SELECT order_id
FROM order_line_t ;
vs.
SELECT DISTINCT order_id    ← Only distinct order numbers are displayed
  FROM order_line_t ;
........
    WHERE state IN ('FL', 'PA', 'NJ')    ← Range search
```

◆ **Use SQL to design query view**

Now we can use SQL to design a view of four-table query for the goods(Pic 6.2).

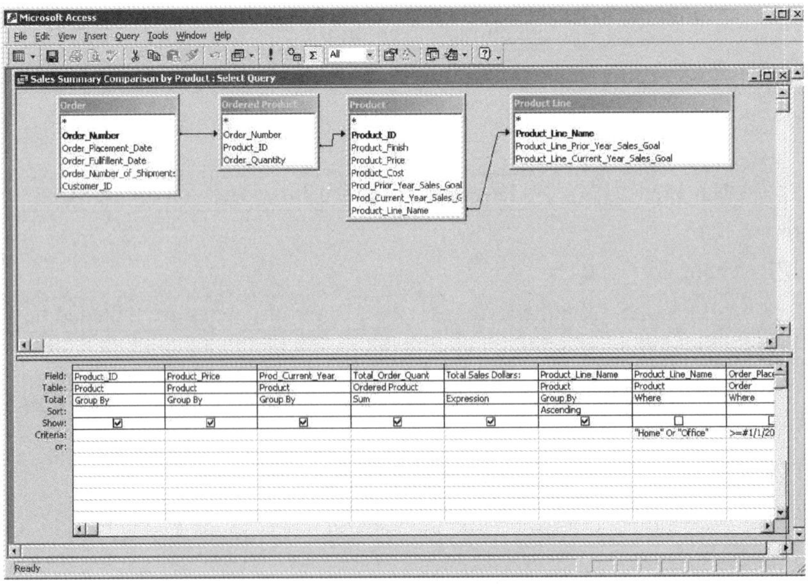

Pic 6.2 A view of four-table query

Now for the SQL View

SELECT Product. Product_ID, Product.Product_Price, Product.Prod_Current_Year_Sales_Goal, Sum([Ordered Product].Order_Quantity) AS Total_Order_Quantity, Sum([Order_Quantity]*[Product_Price]) AS [Total Sales Dollars], Product. Product_Line_Name

FROM (Product RIGHT JOIN ([Order] LEFT JOIN [Ordered Product] ON Order.Order_Number = [Ordered Product].Order_Number) ON Product.Product_ID = [Ordered Product].Product_ID) LEFT JOIN [Product Line] ON Product.Product_Line_Name = [Product Line].Product_Line_Name

WHERE (((Product.Product_Line_Name)="Home" Or (Product.Product_Line_Name)="Office") AND ((Order.Order_Placement_Date)>=#1/1/2002# And (Order.Order_Placement_Date)<#1/1/2003#))

GROUP BY Product.Product_ID, Product.Product_Price, Product.Prod_Current_Year_Sales_Goal, Product.Product_Line_Name

ORDER BY Product.Product_Line_Name;

The result of this query is shown below(Pic 6.3).

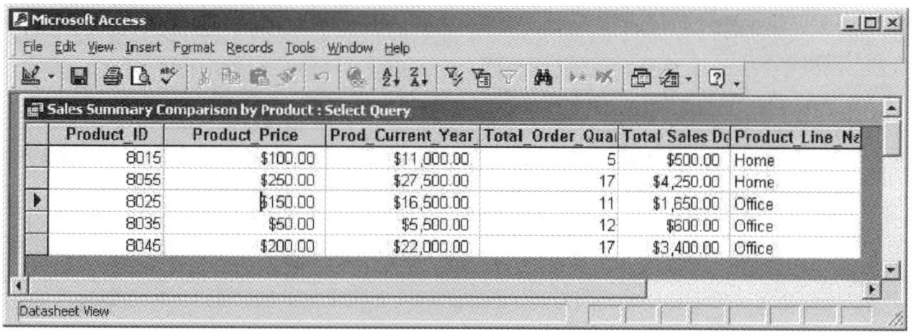

Microsoft Access

Sales Summary Comparison by Product : Select Query

Product_ID	Product_Price	Prod_Current_Year	Total_Order_Qua	Total Sales Do	Product_Line_Na
8015	$100.00	$11,000.00	5	$500.00	Home
8055	$250.00	$27,500.00	17	$4,250.00	Home
8025	$150.00	$16,500.00	11	$1,650.00	Office
8035	$50.00	$5,500.00	12	$600.00	Office
8045	$200.00	$22,000.00	17	$3,400.00	Office

Pic 6.3 The result of the query

The resulting table will contain:

- all columns from each of the source tables
- an instance of each row of the data from each source table

Exercise

Ex 1 Fill in the form according to the six stages of database design.

1	需求分析	
2	概念结构设计	
3	逻辑结构设计	
4	物理结构设计	
5	数据库实施	
6	数据库运行和维护	

Ex 2 Use SQL to create a database for the Product_Finish, the Average Price should be less than $750.

Ex 3 Use SQL to design a view of Comparison of product query.

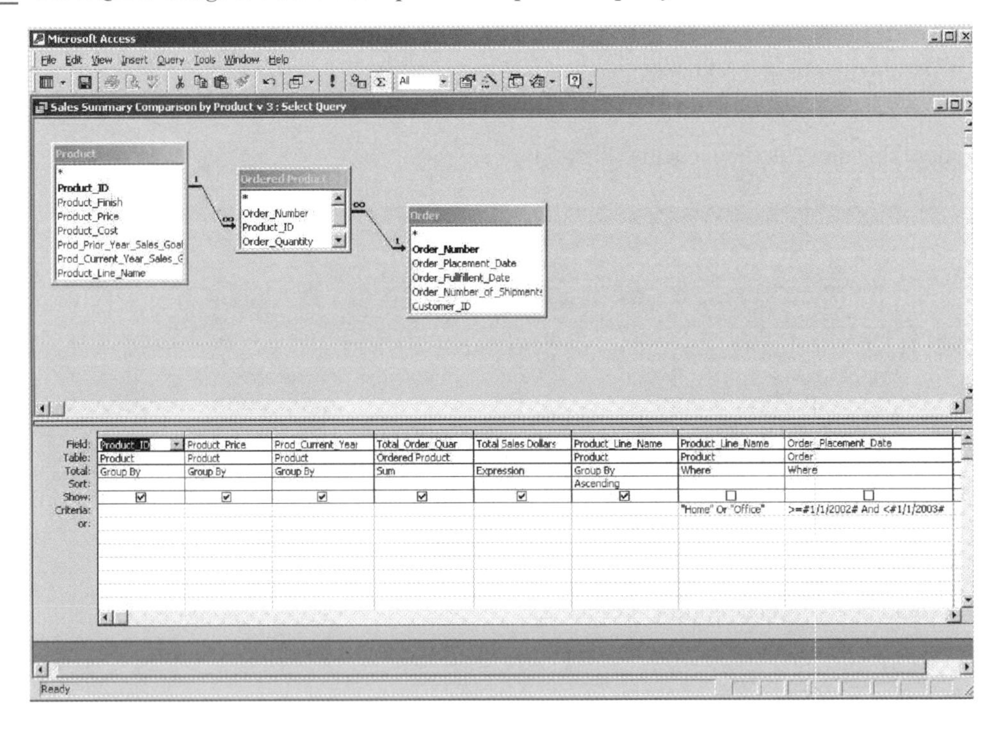

View of the Output of the above query:

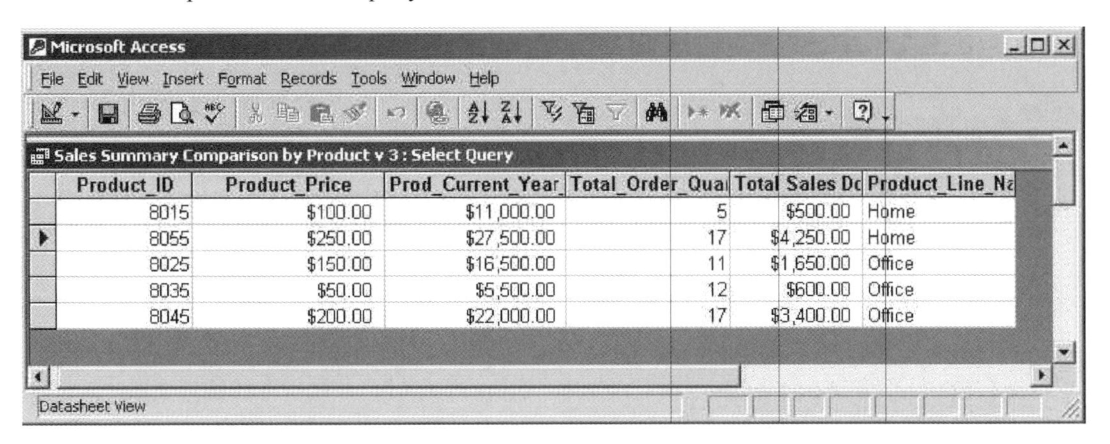

Sales Summary Comparison by Product v 3 : Select Query

Product_ID	Product_Price	Prod_Current_Year	Total_Order_Qua	Total Sales Dc	Product_Line_Na
8015	$100.00	$11,000.00	5	$500.00	Home
8055	$250.00	$27,500.00	17	$4,250.00	Home
8025	$150.00	$16,500.00	11	$1,650.00	Office
8035	$50.00	$5,500.00	12	$600.00	Office
8045	$200.00	$22,000.00	17	$3,400.00	Office

Ex 4 **For the customers, list their IDs and names, and the orders they placed including order number, order date and total value($). Don't show the individual line item for each order. Show orders with a total of over $1000. Finally, sort the output by customer name.**

Part C Occupation English

Training Target

In this part, students are trained to be skilled at giving right instructions to download Microsoft Access on the SQL Server Database and make it compatible and safe.

Making Databases Compatible 使数据库兼容

Post 1: Staff of Compaq Computers Customer Support（岗位1：康柏电脑客户支持）

Post 2: Staff of Microsoft Customer Support（岗位2：微软客户支持）

Post 3: Database Administrator（岗位3：数据库管理员）

A : Compaq Computers Customer Support, what can I do for you?
康柏电脑客户支持，我能为你做些什么？

B : Hello. I am trying to install Microsoft Access① on Windows with a Compaq SQL Server database, but I don't know how to do it. Is there an easy way?
你好。我正试着在装有康柏电脑SQL Server序数据库的系统上安装微软Access程序，但我不知道该怎么做。有没有比较容易的方法？

A : Yes, it can be very easy. Just log onto our website and find the Compaq SQL Connector download page. There you can get an ODBC② driver for a Compaq SQL Sever. Install the Windows version, and you will be able to use Access to retrieve data from Compaq SQL databases.
其实挺简单的。只要登录我们的网站，找到康柏电脑SQL连接器下载页，你就能获得可用于康柏电脑SQL Server的ODBC驱动程序。安装Windows版本，你将可以利用Access检索康柏电脑SQL数据库中的数据。

B : That sounds great! Thank you very much.
很好，谢谢你。

A : You're welcome. Thank you for using Compaq SQL, and have a nice day.
不客气。感谢你使用康柏电脑SQL，祝你度过愉快的一天。

Exercise: Practice the dialogue in pairs.

---

①Access:微软出品的关联式数据库管理系统。它是微软办公系统的成员之一。Access能把数据储存于任何ODBC兼容的数据库内。

②ODBC(Open DataBase Connectivity)开放式数据库连接。它提供了一种标准的应用程序编程接口来访问数据库管理系统。ODBC的设计者们努力使它具有最大的独立性和开放性：与具体的编程语言无关，与具体的数据库系统无关，与具体的操作系统无关。

Word Building

前缀/后缀由一个或几个字母组成，放在词根或单词之前/之后，组成一个新词。

(1) over-(前缀)：过分、在……上面；超过；额外

load：载 ———— overload：超载

head：头 ———— overhead：在头顶上，在上头

(2) tele-(前缀)：电

vision：视力 ———— television：电视

graph：图表 ———— telegraph：电报

(3) re-(前缀)：再、重新

setup：设置 ———— resetup：重新设置

run：运行 ———— rerun：重新运行

(4) -less(后缀)：没有；无；不

use：有用的 ———— useless：无用的

harm：有害的 ———— harmless：无害的

Ex Translate the following words and try your best to guess the meaning of the word on the right according to the clues given on the left.

write	写（动词）	overwritten	________
		rewrite	________
communication	通信（名词）	telecommunication	________
byte	字节(名词)	megabyte	________
		gigabyte	________
relation	联系、关系（名词）	relational	________
manipulate	处理、操作（动词）	manipulation	________
character	特点（名词）	characteristic	________
wire	线（名词）	wireless	________
server	服务器（名词）	service	________

Ex 1 **What's a standardized query language for requesting information from a database?**

Ex 2 **Fill in the table below by matching the corresponding Chinese or English equivalents.**

SQL	
	客户/服务器
DML	
	交互界面
DDL	
	检索

Ex 3 **Choose the best answer to the following questions according to the text we learnt.**

1. The original version called SEQUEL was designed by an ______ research center in 1974 and 1975.
   A. IBM
   B. ISO
   C. Intel
   D. ANSI
2. SQL's ______ commands make it easy for you to design and manage the information in your database.
   A. long
   B. short
   C. simple
   D. complicated
3. The most commonly used SQL commands are classified as________ language commands.
   A. DML, DCL and LCD
   B. DDL, LCD and DCL
   C. DDL, DML and LCD
   D. DDL, DML and DCL
4. The _______ command is the most commonly used command in SQL.
   A. INSERT
   B. SELECT
   C. WHERE
   D. CREATE
5. You can use_________ command to insert, update, delete, and retrieve information from the tables in a database.
   A. DDL
   B. DML
   C. DCL
   D. SELECT

Project Seven

Online Shopping Mall's Security

Part A Theoretical Learning

Part B Practical Learning

Part C Occupation English

Part A Theoretical Learning

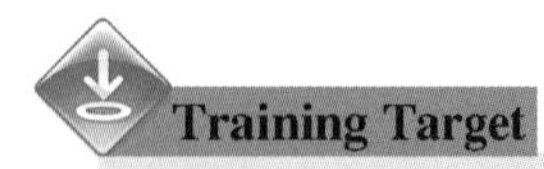

Training Target

In this part, our target is to improve the speed of reading professional articles and the comprehension ability of the reader. We have marked specialized vocabulary key words in some paragraphs so that the reader can quickly grasp the main idea of the sentences and paragraphs.

Skill One Computer Virus

What is computer virus? **A computer virus is a special kind of computer program that reproduces its own code by attaching itself to other executable files, and spreads usually across disks and networks surreptitiously.**

Viruses have many different forms, but they all potentially have two phases to their execution: the **infection phase** and the **attack phase**.

◇ **Infection Phase**

The virus has the potential to infect other programs when it executes. We often don't clearly understand when the virus will infect the other programs. Some viruses infect other programs each time they are executed; other viruses infect only upon a certain **trigger**. This trigger could be anything: a day or a time, an external event on your PC, a counter within the virus, etc. Virus writers want their programs to spread as far as possible before anyone notices them.

It is a serious mistake to execute a program a few times—find nothing infected and presume there are no viruses in the program. Maybe the virus simply hasn't yet triggered its infection phase!

Many viruses go resident in the memory of your PC in the same or similar way as terminate and stay resident(**TSR**) programs. This means the virus can wait for some external event before it infects additional programs. The virus may silently **lurk** in memory waiting for you to access a diskette, copy a file, or execute a program before it infects anything. This makes viruses more difficult to analyze since it's hard to guess what trigger condition they use for their infection.

virus ['vairəs] n. 病毒

surreptitiously [ˌsʌrəp'tiʃəsli] adv. 秘密地

infection phase 感染期

attack phase 攻击期

infect [in'fekt] vt. 传染，感染

trigger ['trigə] n. 触发事件

TSR abbr. 终止驻留程序

lurk [lə:k] v. 潜伏

Resident viruses frequently take over portions of the system software on the PC to hide their existence. This technique is called stealth. **Polymorphic** techniques also help viruses to infect yet avoid detection.

resident virus 驻留的病毒
polymorphic [pɔli'mɔ:fik] adj. 多态的

Note that **worms** often take the opposite approach to spread as fast as possible. While this makes their detection virtually certain, they also have the effect of bringing down networks and denying access to the network. This is one of the goals of many worms.

worm [wə:m] n. 蠕虫

During the infection phase, in order to infect a computer, a virus has to have the chance to execute its code. Viruses usually ensure that this happens by behaving like a **parasite**, that is, by modifying another items so that the virus codes are executed when the legitimate items are run or opened.

parasite ['pærəsait] n. 寄生虫

Good vehicles for viruses include the parts of a disk which contain codes executed. As long as the virus is active on the computer, it can copy itself to files or disks that are accessed.

Viruses can be transmitted by:

△ Booting a PC from an infected medium
△ Executing an infected program
△ Opening an infected file

Common routes for virus **infiltration** include:

△ Floppy disks or other media that users can exchange
△ **E-mail attachment**
△ **Pirated software**
△ Shareware

infiltration [ˌinfil'treiʃən] n. 渗入
E-mail attachment 邮件附件
pirated software 盗版软件

◇ **Attack Phase**

Many viruses do unpleasant things such as deleting files or changing random data on your disk or merely slowing your PC down; some viruses do less harmful things such as playing music or creating messages or animation on your screen. Just as the infection phase can be triggered by some events, the attack phase also has its own trigger.

delete file 删除文件

Does this mean a virus without an attack phase is **benign**? No. Most viruses have **bugs** in them and these bugs often cause unintended negative side effects. In addition, even if the virus is perfect, it still steals system resources.

benign [bi'nain] adj. 良性的
bug[bʌg] n. 漏洞

Viruses often delay revealing their presence by launching their attack only after they have had **ample** opportunity to spread. This means the attack could be delayed for days, weeks, months, or even years after the first **initial** infection.

ample ['æmpl] adj. 充足的
initial [i'niʃəl] adj. 开始的，最初的

The attack phase is optional, many viruses simply reproduce and have no trigger for an attack phase. Does this mean these are "good" viruses? No! Anything that writes itself to your disk without your permission is stealing storage and CPU **cycles**. What is worse is that viruses of "just infect" without an attack phase, often damage the program or disks they infect. This is not an intentional act of virus, but simply a result of the fact that many viruses contain extremely poor defective codes.

cycle ['saikl] n. 周期

As an example, one of the most common viruses, stoned, is not intentionally harmful. Unfortunately, the author did not anticipate the use of anything other than 360K floppy disks. The original virus tried to hide its own code in an area of 1.2MB that resulted in corruption of the entire diskette.

.End.

Key Words

virus n.病毒
lurk v.潜伏
cycle n.周期
infiltration n.渗入
bug n.漏洞
trigger n.触发事件
worm n.蠕虫
parasite n.寄生虫
benign adj.良性的
pirated software 盗版软件
E-mail attachment 邮件附件
attack phase 攻击期
delete file 删除文件
infection phase 感染期
resident virus 驻留的病毒
TSR (Terminate and Stay Resident) abbr.终止驻留程序

参考译文 技能1 计算机病毒

什么是计算机病毒？计算机病毒是以把自身附加到可执行文件的方式来复制其自身代码的特殊的计算机程序，并常常利用磁盘和网络秘密地进行传播。

病毒具有多种不同的形式，但是它们在执行中都要有两个时期：感染期和攻击期。

◇ 感染期

病毒在执行时可能会感染其他程序。我们通常无法清楚了解病毒究竟要在何时感染其他程序。一些病毒在每一次执行时都会感染其他程序，另一些病毒只有当某一个触发事件发生时才能进行传染。这种触发事件可能是任何事件：一个日期或某一个时间，一个计算机的外部事件，一个病毒内部的计数器等。病毒的编写者希望他的程序能在其他人发现之前尽可能地广泛传播。

对一个程序执行了几次，发现没有受到感染，就认为程序里没有病毒，这种想法是一种严重的错误。也许病毒还没有进入它的感染期！

许多病毒驻存在电脑的内存中，其方式与终止驻留程序相同或相似。这就意味着病毒可以等待某个外部事件的发生，然后再感染其他程序。病毒可以默默地潜伏在内存中，等你访问一个磁盘、复制一个文件或执行一个程序之后，才开始进行感染。这使病毒更加难以分析，因为很难判断病毒要使用什么样的触发条件来进行感染。

驻留的病毒经常利用部分电脑系统软件来隐藏它们的存在。这种技术被称为隐蔽。多形态技术也有助于病毒进行感染而不被发现。

需要注意的是蠕虫病毒经常采用相反的方法尽可能快地进行传播。尽管一定能够发现它们，但该病毒在被发现前已经导致了网络速度的下降，并阻止了对网络的访问。这是许多蠕虫病毒的目标之一。

在感染期，为了要传染某台计算机，病毒必须得有运行其代码的机会。为了运行代码，病毒通常会像寄生虫一样，通过修改其他程序，在合法程序运行或打开时运行病毒代码。

实现病毒传染的常见工具包括可执行代码的部分磁盘，只要病毒在计算机上是活跃的，它就能将其自身复制到计算机所访问的其他文件或磁盘上。

病毒可通过以下方式传输：

△ 通过已被感染的媒体启动计算机

△ 运行被传染的程序

△ 打开被传染的文件

病毒渗入的常见途径包括：

△ 用户间能交换的盘片或其他媒体

△ 邮件附件

△ 盗版软件

△ 共享软件

◇ 攻击期

许多病毒做一些令人十分不愉快的事情，如删除文件，或任意修改盘上的数据，或仅仅使电脑的运行速度变慢；某些病毒做一些危害性较小的事情，如播放音乐或在屏幕上显示信息或动画。就像需要某个事件来触发感染期一样，攻击期也需要自己的触发事件。

这是否意味着没有攻击阶段的病毒就是良性的呢？非也！大多数病毒都有漏洞，这些漏洞经常会引起负面效果。另外，即使病毒没有漏洞，它仍然会窃取系统资源。

病毒经常在有充分的机会来进行传播后，才进行攻击，这样就可以延迟病毒的暴露。这意味着病毒在第一次感染后，可能要延迟几天、几周、几个月甚至几年才进行攻击。

攻击期是随意的，许多病毒仅仅是繁殖，而没有攻击期的触发事件。这是否表明这些病毒就是“好”病毒呢？非也！任何未经允许就将自己写入磁盘的病毒，都在窃取存储空间和CPU周期。更糟的是，这些“仅仅感染”而没有攻击期的病毒经常破坏它们所感染的程序或磁盘。这不是病毒的故意行为，而仅仅是因为许多病毒包含了有缺陷的代码。

例如，stoned是过去常见的病毒之一，该病毒并不是故意要具有危害性的。但不幸的是，编写者并没有预见到不同于360K软盘的其他磁盘的使用。最初病毒只试图在1.2M字节的区域上隐藏自己的代码，结果却破坏了整个磁盘。

Skill Two Computer System Security Measures

There is an urgent need for computer security. Computer owners must take measures to prevent theft and **inappropriate** use of their equipment. One aspect of computer security is the protection of information against inappropriate manipulation, destruction or **disclosure**. Another security problem concerns the protection of the computer system and data on the computer. It is essential that security measures protect all operating systems.

Computer security is concerned with protecting information, hardware, and software. Security measures consist of **encryption**, **restricting access**, **data layering**, and **backing up data**.

◇ **Encryption**

Encryption provides secrecy for data. Since data of encryption that cannot be read generally and also cannot be changed, encryption can be used to achieve **integrity**. Encryption is an important tool in computer security, but one should not **overrate** its importance. Users must understand that encryption does not solve all computer security problems. Furthermore, if encryption is not used properly, it can have no effect on security or can degrade the performance of the entire system. Thus, it is important to know the situations in which encryption is useful and how to use it effectively.

The most powerful tool in providing computer security is coding. It is **unintelligible** for the outside observer to transform data, the value of an **interception** and the possibility of a **modification** or a **fabrication** are almost **mollified**.

◇ **Restricting Access**

Everyone in the organization needs to access to the Internet. Employees should be trained on the threats that computer viruses can **impose on** information systems. They should realize how important it is not to open unidentified files, leave computer terminals on while they are away from their desks, display their passwords for everyone to see, etc.

Security experts are constantly **devising** new ways to protect computer systems from access by unauthorized persons. Sometimes security is a matter of placing guards in company computer rooms and checking the **identification** of everyone admitted. The passwords are secret words or numbers that must be keyed into a computer system to gain access.

inappropriate [inə'prəupriət] adj. 不适当的

disclosure [dis'kləuʒə] n. 泄露

encryption [in'kripʃən] n. 加密

restricting access 限制访问

data layering 数据分层

backing up data 备份数据

integrity [in'tegriti] n. 完整

overrate [ˌəuvə'reit] v. 过高评价

unintelligible ['ʌnin'telidʒəbl] adj. 难以理解的

interception [ˌintə'sepʃən] n. 截取

modification [ˌmɔdifi'keiʃən] n. 更改，修改

fabrication [ˌfæbri'keiʃən] n. 伪造

mollify ['mɔlifai] v. 安慰

impose on 占……便宜；利用

devise [di'vaiz] v. 设计

identification [aiˌdentifi'keiʃən] n. 身份证明

oftentimes ['ɔfəntaimz] adv. 屡次，时常地

Most major corporations today use special hardware and software called **firewalls** which act as a security **buffer** between the corporation's private network and all external networks, including the Internet to control the computer network access. All electronic communications coming into and leaving the corporation must be evaluated by the firewall. Security is maintained by **denying** access to unauthorized communications.

firewall ['faiəwɔ:l] n. 防火墙
buffer ['bʌfə] n. 缓冲，缓冲区
deny [di'nai] v. 否认，拒绝

◇ **Data layering**

Organizations need to determine and classify important information, such as the company's financial information, employee data, etc. and establish different layers of security. Only certain employees should have access to this classified information.

◇ **Backing up data**

Equipment can always be replaced. A company's data, however, may be **irreplaceable**. Most companies should take some effective measures of trying to keep software and data from being tampered with in the first place. These measures include careful screening job applicants, guarding passwords, and **auditing** data and programs from time to time. The safest procedure, however, is to make data backups frequently and to store them in remote locations.

irreplaceable [iri'pleisəbl] adj. 不可替代的
audit ['ɔ:dit] v. 审计

From this discussion, it should be evident that how important computer security is. Computer security will continue to be a problem we need to focus on because the number of computers and users continues to grow.

End.

Key Words

encryption n. 加密	interception n. 截取
fabrication n. 伪造	integrity n. 完整
firewall n. 防火墙	buffer n. 缓冲，缓冲区
restricting access 限制访问	data layering 数据分层
modification n. 更改，修改	backing up data 备份数据

参考译文 技能2 计算机系统安全措施

计算机安全已成为急需解决的问题。计算机拥有者必须采取措施防止他们的设备被窃取和非法使用。计算机安全的一个方面就是要保护信息不被越权操作、破坏或泄露。另一个安全问题涉及计算机上操作系统和数据的保护。采用安全措施保护所有的计算机操作系统是很有必要的。

计算机安全与信息、硬件和软件的保护有关。安全措施包括加密、限制访问、数据分层和备份数据。

◇ 加密

加密提供数据保密。因为加密的数据一般不能读出，也不能更改，因此能保证数据的完整。加密是计算机安全的重要工具，但有时也不能对它的重要性估计过高。用户应该知道加密并不能解决计算机所有的安全问题。甚至如果加密使用不当，不但对安全没有作用，还会降低整个系统的性能。所以，重要的是要了解在什么情况下加密有用和有效。

保证计算机安全的最有效的方法是编码。对外界而言，转化数据是无规律的，这样截取的数据就是没有用处的，修改或伪造的可能性就不会发生。

◇ 限制访问

公司内的每个人都需要访问因特网。每位职员都应在计算机病毒可利用信息系统造成威胁这一方面进行培训。他们应该明白不要打开未经确认的文件，当他们离开办公桌时不要让电脑终端继续工作，不要让其他任何人看到自己的密码等，这些都是很重要的。

安全专家不断设计新方法，用以保护计算机系统免受未经授权的人的访问。有时，为保证安全，会派警卫看护公司计算机室，检查每个进入的人的身份证明。口令是秘密的单词或数字，必须将其键入计算机系统才能进行访问。

今天，大多数大公司都使用被称为防火墙的专门的硬件和软件来控制计算机网络的访问。这些防火墙，在公司专用网络与包括因特网在内的所有外部网络之间，起到安全缓冲区的作用。所有进出公司的电子通信都必须经过防火墙的评估。通过拒绝未经授权的通信进出来维护公司的网络安全。

◇ 数据分层

公司必须确定哪些是重要信息并将其进行分类。比如公司的财政信息、职员数据等，并建立不同的安全层次。只有特定的职员能访问这些分类信息。

◇ 备份数据

设备随时可以替换。然而，一个公司的数据可能是无法替代的。因此，大多数公司首先会采取一些有效的措施，防止软件和数据被篡改。这些方法包括仔细审查求职者、严守口令，以及时常检查数据和程序。然而，最保险的办法是经常制作数据备份，并将其存放在远距离地点。

通过以上讨论，我们应当清楚计算机安全的重要性。随着计算机和计算机用户人数的不断增加，计算机安全将会继续成为需要人们关注的问题。

Fast Reading One | A Brief Introduction to Firewall

Firewalls are frequently used to prevent unauthorized Internet users from accessing private networks connected to the Internet. A small home network has many of the same security issues that a large corporate network does. You can use a firewall to protect your home network and family from offensive websites and potential hackers. Firewalls can be implemented in both hardware and software, or a combination of both. A firewall is simply a program or hardware device that filters the information coming through the Internet connection into your private network or computer system. All messages entering or leaving the Internet will pass through the firewall, which examines each message and blocks those that do not meet the specified security criteria. If an incoming packet of information is flagged by the filters, it is not allowed to pass through.

Typically, a firewall is placed on the entry point to a public network such as the Internet. It is a protective system that lies between your computer and the Internet. It should be considered as a traffic cop. The firewall's role is to ensure that all communication between an organization's network and the Internet conforms to the organization's security policies. Its job is similar to a physical firewall that keeps a fire from spreading from one area to the next. Primarily these systems are TCP/IP-based, and depending on the implementation, can enforce security roadblocks as well as provide administrators with answers to the following questions:

△ Who's been using network?

△ What were they doing on network?

△ When were they using network?

△ Where were they going on network?

△ Who failed to enter the network?

End.

参考译文 | 防火墙简介

防火墙经常用于防止未经许可的因特网用户访问连接到因特网上的个人网络。小型家庭网有着与大公司的网络相同的安全问题。防火墙可以保护你的家庭网和家庭免遭恶意网站和潜在黑客的攻击。防火墙可以由硬件、软件或二者联合实现。一个防火墙就是一个程序或者一台硬件设备，用于过滤通过因特网连接进入你的专用网或计算机系统中的信息。进入或由内部网发出的所有信息都要经过防火墙，它检查每一个不符合指定安全标准的信息。如果一个输入的信息包被过滤器做了标记，它就不允许通过。

典型的防火墙置于公共网络（如因特网）入口处。它是位于你的计算机与因特网之间的保护系统。它可以被看作是交通警察。防火墙的作用是确保一个单位的网络与因特网之间的所有通信都符合单位的安全策略。它的作用类似于一堵防止火灾从一处蔓延到另一处的实实在在的防火墙。这些系统基本上遵循TCP/IP协议，并与实现方法有关，它们能实施安全路障并为管理人员提供下列问题的答案：

△ 谁在使用网络?

△ 他们在网上做什么?

△ 他们什么时间使用网络?

△ 他们去了网络的什么地方?

△ 谁要上网但没有成功?

Fast Reading Two Firewall Techniques

In general, there are three types of firewall implementations, some of which can be used together to create a more secure environment. These implementations are packet filtering, application proxies, and circuit-level or generic-application proxies. Packet filtering is often achieved in the router itself. Application proxies, on the other hand, usually run on standalone servers. Proxy services take a different approach than packet filtering, using a modified client program that connects to a special intermediate host that actually connects to the desired service.

There are several types of firewall techniques:

◇ **Application gateway**

The first firewalls are application gateways, and are sometimes known as proxy gateways. These are made up of the bastion hosts that run special software to act as a proxy server. This software runs at the Application Layer of the OSI Reference Model applies security mechanisms to specific applications, such as FTP and Telnet servers. This is very effective.

◇ **Packet filter**

Looks at each packet through the network and accepts or rejects it based on user-defined rules. Packet filtering is fairly effective and transparent to users, but it is difficult to configure. In addition, it is susceptible to IP spoofing.

◇ **Circuit-level gateway**

Applies security mechanisms when a TCP or UDP connection is established. Once the connection has been made, packets can flow between the hosts without further checking.

◇ **Proxy server**

Intercepts all messages entering and leaving the network. The proxy server effectively hides the true network addresses.

Firewalls use one or more of three methods to control traffic flowing in and out of the network:

(1) Packet filtering: Packets (small chunks of data) are analyzed against a set of filters. Packets that make it through the filters are sent to the requesting system and all others are discarded.

(2) Proxy service: Information from the Internet is retrieved by the firewall and then sent to the requesting system and vice versa.

(3) Stateful inspection: A newer method that doesn't examine the content of each packet but instead compares certain key parts of the packet to a database of trusted information. Information traveling from inside the firewall to the outside is monitored for specific defining characteristics, then

incoming information is compared to these characteristics. If the comparison yields a reasonable match, the information is allowed to pass through. Otherwise it is discarded.

The level of security you establish will determine how many of these threats can be stopped by your firewall. The highest level of security would be to simply block everything. Obviously that defeats the purpose of having an Internet connection. But a common rule is to block everything, then begin to select what types of traffic you will allow. You can also restrict traffic that travels through the firewall so that only certain types of information, such as E-mail, can get through. For most of us, it is probably better to work with the defaults provided by the firewall developer unless there is a specific reason to change it.

One of the best things about a firewall from a security standpoint is that it stops anyone on the outside from logging onto a computer in your private network. While this is a big idea for businesses, most home networks will probably not be threatened in this manner.

.End.

参考译文 防火墙技术

通常有三种类型的防火墙实现方案，有些是将几种类型一起使用，以确保建立一个更安全的环境。这些实现方案是：数据包过滤、应用程序代理和电路级或通用应用程序代理。数据包过滤常常是在路由器中实现的，而应用程序代理通常运行在独立的服务器上。代理服务采取不同于数据包过滤的方法，使用修改的客户端程序，该程序与专用中间主机相连，而该主机又与实际所需服务器端相连。

这里有几种不同类型的防火墙技术：

◇ **应用网关**

第一种防火墙是应用网关，即平时所知的代理网关。应用网关由运行特定软件的堡垒型主机组成，作用如同代理服务器。软件运行在OSI参考模型的应用层上，并将安全机制应用于专用应用软件，比如FTP与Telnet服务器上，这是非常有效的。

◇ **数据包过滤器**

按照用户定义的规则，对通过网络的每一个数据包进行检查并接收或拒绝。数据包过滤是非常有效和对用户透明的，但是设置困难。此外，它易受IP的欺骗。

◇ **电路级网关**

当TCP或者UDP建立连接时，电路级网关会开启安全机制。一旦建立了连接，数据包可以在主机间流动且不需要进一步检查。

◇ **代理服务器**

代理服务器截取所有进入网络和由网络发出的信息。它能有效地屏蔽真实的网络地址。

防火墙使用下列三种方法中的一种或几种来控制进出网络的通信：

（1）数据包过滤：数据包（小块数据）由一组过滤器进行分析。能通过过滤器的数据包被发送到发出请求的系统，其他的则被丢弃。

（2）代理服务：来自因特网的信息通过防火墙进行检索，然后发送到提出请求的系统，反之亦然。

（3）状态检查：一种更新的方法，并不检查每个数据包的内容，而是将数据包的某个关键部分与一个可信的信息数据库进行比较。从防火墙内部传输到外部的信息可根据特别规定的特性进行监控，然后系统会将输入的信息与这些特性相比较，若能生成一个合理的匹配，则允许信息通过，若不能就会丢弃信息。

你所设定的安全级别将决定你的防火墙能阻挡多少威胁。最高安全级别就是阻断一切。很显然，这就失去了进行互联网连接的意义。但通常的做法是阻断一切，然后，开始选择你将允许什么样的流量类型。你还可以限制通过防火墙的流量，以便只有几种信息可以通过，如电子邮件。对我们大多数人来说，除非有特殊的理由要改变它，否则最好在防火墙开发商提供的默认条件下工作。

从安全的角度来看，防火墙的一个优点就是它能阻止任何外来人员登录到专用网中的计算机上，这一点对企很重要，大多数家庭网不受这种方式的威胁。

Ex 1 **What is a computer virus? Try to describe it in some words.**

Ex 2 **How does a computer virus infect files?**

Ex 3 **Fill in the table below by matching the corresponding Chinese or English equivalents.**

resident virus	
	病毒
worm	
	攻击期
bug	
	感染期

Ex 4 Choose the best answer to the following questions according to the text we learnt.

1. A virus is a________.

   A. program

   B. computer

   C. bad man

   D. beast

2. A virus is a program that reproduces its own code by________.(多选)

   A. adding to the end of a file

   B. inserting into the middle of a file

   C. simply placing a pointer

   D. replacing another program

3. Similar to viruses, you can also find malicious code in ________.(多选)

   A. Trojan Horses

   B. Worms

   C. Microsoft Word Documents

   D. Logic bombs

4. Viruses all have two phases to their execution, the ________and the ________.(多选)

   A. infection phase

   B. deletion phase

   C. attack phase

   D. creation phase

5. ________ maybe a certain cause that some viruses infect upon.(多选)

   A. A day

   B. A time

   C. An external event on your PC

   D. A counter within the virus

6. Many viruses go resident in the memory like a ________.

   A. exe file

   B. com file

   C. TSR program

   D. data file

7. If you ________, it may wake up a virus that is resident in memory. (多选)

   A. delete a file

   B. access a diskette

   C. execute a program

   D. copy a file

8. Viruses can be transmitted by ________.

   A. booting a PC from an infected medium

   B. executing an infected program

   C. opening an infected file

   D. all of the above

9. Common routes for virus infiltration include ________.

   A. floppy disks or other media that users can exchange

   B. E-mail attachments

   C. pirated software and shareware

   D. All of the above

10. If a virus simply reproduces and has no cause for an attack phase, but it will still _______ without your permission.

   A. hide its own code

   B. steal storage and pilfer CPU cycles

   C. delete files

   D. play music

Part B Practical Learning

Training Target

In this part, students must finish two special tasks in English environment, under the guidance of the specialized English teacher. The students must work with each other in the same group.

Task One Discuss Potential Security Issues

In this task, students should understand the problems about the security of the online shop.

Usually there are some aspects about the security of the online shop: First website is easy to be infected with virus, secondly, website is regularly attacked by hackers. For example, customers' private information is intercepted and stolen; order information is revised; the online store's information and the customer information are fake. Other problems may include online payment security, payment repudiation and so on.

Task Two Set up Security Measures for the Store

In this task, students can build safety measures to cope with the common online shop security issues.

These measures include:

Install anti-virus software on your computer and keep it updated;

Strengthen the personal identity verification;

Encrypt the purchase process data;

Perfect the methods of online payment;

Build online store real name system and credit system.

Part C Occupation English

Training Target

In this part, students are supposed to practice the dialogue, who act as staff of service center. Offering help to the customers in operation system installations is one of the common tasks in their future job.

Web Security 网络安全

Post 1: Staff of Kingsoft Online Technical Support Center（岗位：金山在线技术支持）

Post 2: Network Engineer of Shanghai Stock Exchange Firm（岗位：上海证券公司工程师）

A : Kingsoft Online Technical Support Center. Can I help you?
金山在线技术支持中心，我能帮您做什么？

B : Hi, I'm from Shanghai Stock Exchange Firm. I want to ask how to enhance our corporate network security. Our company has used your enterprise anti-virus and firewall products for a while. But we are not very satisfied with that. Are there any other products to ensure better security?
你好，这里是上海证券公司。我想了解一下如何加强我们公司的网络安全。我们公司使用贵公司的企业版防病毒和防火墙产品已经有一段时间了，但我们不是很满意。有没有其他更能保障网络安全的产品呢？

A : There are several ways to enhance your network security; adding an anti-spyware program to the anti-virus program could be a good choice. Currently we have added a new anti-spyware module to General Computers Virus Scan Enterprise. You should try that.
有好多种方法可以加强贵公司的网络安全，在防病毒程序的基础上，再安装一个反间谍软件会是一个非常不错的选择。目前我们已经给通用电脑病毒扫描企业版添加了一个新的反间谍模块。您不妨试一下。

B : That sounds good. I will check about that. Could you send me a brochure of that product?
听起来不错。我想试一下。你能寄给我一份产品说明书吗？

A : Sure, and your customer ID, please?
当然可以，请告诉我您的客户ID号。

B : Err, hold on a second. OK, I got it. It's TA2816.
稍等一下。哦，找到了。是TA2816.

A : OK, I got it. You will be receiving our brochure within a week.
好的，您将在一周之内收到我们的产品说明书。

B : Thank you very much. 太感谢你了。

A : You're welcome. Thank you for your interest in our products and services.
不客气。感谢您关注我们的产品和服务。

Word Building

前缀/后缀由一个或几个字母组成，放在词根或单词之前/之后，组成一个新词。

(1) micro-(前缀)：微，小

processor：处理器 ———— microprocessor：微处理器

(2) super-（前缀）：超

market ：市场 ———— supermarket ：超级市场

highway ：公路 ———— superhighway ：超级公路

(3) -er （后缀)：……者，……物，……的人

teach：教 ———— teacher：教师、导师

cook ：烹调，煮 ———— cooker：厨具

(4) -able（后缀）：能……的，可……的，易于……的

adjust：调整 ———— adjustable：可调整的

suit ：适合 ———— suitable：适当的，合适的

Ex Translate the following words and try your best to guess the meaning of the word on the right according to the clues given on the left.

computer	计算机（名词）	supercomputer ____________
computer	计算机（名词）	microcomputer ____________
electronic	电子的（形容词）	electron ____________
		electronics ____________
number	数字 （名词）	numerical ____________
add	加（动词）	adder ____________
multiply	乘 （动词）	multiplier ____________
divide	分，除（动词）	divider ____________
flex	折曲 （动词）	flexible ____________
rely	依靠（动词）	reliable ____________

Exercise

Ex 1 According to the text, if you have a system that is not currently running virus protection software, what should you do?

Ex 2 Fill in the table below by matching the corresponding Chinese or English equivalents.

	防火墙
data layering	
	缓冲
modification	
	备份数据
restricting access	
	截取

Ex 3 Choose the best answer to the following questions according to the text we learnt.

1. Security measures consist of __________.
   A. encryption
   B. restricting access
   C. data layering and backing up data
   D. all of the above
2. The most powerful tool in providing computer security is __________.
   A. transforming data
   B. restricting access
   C. coding
   D. backing up data
3. __________ can be used to achieve integrity and secrecy for data.
   A. Restricting access
   B. Encryption
   C. Data layering
   D. Backing up data
4. It is very important that we should not __________ in protecting computer system.
   A. display our passwords for everyone to see
   B. open unidentified files
   C. leave computer terminals on while we are away from our desks
   D. all of the above
5. All electronic communications coming into and leaving the corporation must be evaluated by the __________.
   A. passwords
   B. firewall
   C. gateway
   D. buffer

Project Eight

Letting Students Trade Online

Part A Theoretical Learning

Part B Practical Learning

Part C Occupation English

Part A Theoretical Learning

Training Target

In this part, our target is to improve the speed of reading professional articles and the comprehension ability of the reader. We have marked specialized vocabulary key words in some paragraphs so that the reader can quickly grasp the main idea of the sentences and paragraphs.

Skill One E-Commerce

In the 21st century, the rapid development of information technology and the rapid increase in information exchange have brought new drives and **innovative** ideas to the whole society. The wide adoption of information technology by the community has led to great changes. These changes are not simply in the context of data processing or computing. They are changes which affect how we communicate with each other, how we organize our daily activities, how we educate the younger generation, and how we run business. The development and wide adoption of information technology, computer network and the Internet have **transformed** the mode of operation of many businesses, and at the same time have brought along **unprecedented** business opportunities. Businesses are now able to conduct transactions across geographical **boundaries**, across time zones and at a high efficiency. **E-Commerce** has become the market trend of the Century.

innovative ['ɪnəveitɪv] adj. 革新的；创新的

transform [træns'fɔ:m] vt. 改变；改观

unprecedented [ʌn'presidentid] adj. 空前的，无前例的

boundary['baundəri] n. 分界线

E-commerce 电子商务

What's E-Commerce?

E-Commerce is doing business through electronic media. It means using simple, fast and low-cost electronic communications to transact, without face to face meeting between the two parties of the transaction. Now, it is mainly done through the Internet and Electronic Data Interchange (EDI). E-Commerce was first developed in the 1960s. With the wide use of computer, the maturity and the wide adoption of the Internet, the **permeation** of credit cards, the establishment of secure transaction agreement and the support and

permeation [ˌpə:mi'eiʃn] n. 渗入，透过

promotion by governments, the development of E-Commerce is becoming **prosperous**, with people starting to use electronic means as the media of doing business.

prosperous ['prɔspərəs] adj. 繁荣的，兴旺的

Types of E-Commerce

◇ Electronic network within the company: through the Internet, people can exchange and handle business information internally.

◇ Business-to-Business (**BTB** or B2B) E-Commerce: amongst all other types of E-Commerce, this way of doing business electronically through the Internet or Electronic Data Interchange is the one that deserves the most attention. As estimated by Forrester Research, BTB E-Commerce will grow at a rate three times that of Business-to-Consumer E-Commerce and thus has the greatest potential for growth.

BTB 企业对企业

◇ Business-to-Consumer(**BTC** or B2C): businesses provide consumers with online shopping through the Internet, allowing consumers to shop and pay their bills online. This type of offering saves time for both **retailers** and consumers.

BTC 企业对消费者

retailer['ri:teilə] n. 零售商，零售店

◇ Consumer-to-Consumer: consumers can post their own products online through some agent websites for other consumers to bid.

◇ Government-to-Business: this mode of trading often describes the way in which government purchases goods and services through electronic media such as the Internet, for example the Electronic Tendering System (ETS). This system is an **infrastructure** to provide online services such as registration of suppliers, tender **notification**, downloading facility for tender documents, enquiries handling, submission of tender proposals and announcement of tender results.

infrastructure['ɪnfrəstrʌktʃə] n. 基础设施

notification [ˌnəutɪfɪ'keiʃn] n. 通知，通告

Global trends

In 2010, the United Kingdom had the biggest E-Commerce market in the world when measured by the amount spent per capita. The Czech Republic is the European country where E-Commerce delivers the biggest contribution to the enterprises' total revenue. Almost a quarter (24%) of the country's total turnover is generated via the **online channel**.

online channel 在线渠道

Among emerging economies, China's E-Commerce presence continues to expand. With 384 million Internet users, China's online shopping sales rose to $36.6 billion in 2009 and one of the reasons behind the huge growth has been the improved trust level for shoppers. The Chinese retailers have been able to help consumers feel more comfortable shopping online. China's cross-border E-Commerce is also growing rapidly. E-Commerce transactions between China and other countries increased 32% to 2.3 trillion yuan ($375.8 billion) in 2012 and accounted for 9.6% of China's total **international trade**.

international trade 国际贸易

Other BRIC countries are witnessing the accelerated growth of E-Commerce as well. In Russia, the total E-Commerce market is projected to total somewhere between 690 billion rubles ($23 billion) and 900 billion rubles ($30 billion) in 2015, at 2010 values. This will equal 5% of total retail volume in Russia. Longer-term, the market size of Russian E-Commerce could reach $50 billion by 2020. Brazil's E-Commerce is growing quickly with retail E-Commerce sales expected to grow at a healthy double-digit pace through 2014. By 2016, eMarketer expects retail E-Commerce sales in Brazil to reach $17.3 billion.

India's E-Commerce growth, on the other hand, has been slower although the country's potential remains solid considering its surging economy, the rapid growth of the Internet penetration, English language proficiency and a vast market of 1.2 billion consumers (although perhaps only 50 million access the Internet through PCs and some estimate the most active group of E-Commerce customers numbers only 2~3 million). E-Commerce traffic grew about 50% from 2011 to 2012, from 26.1 million to 37.5 million, according to a report released by Com Score. Still much of the estimated 14 billion dollars in 2012 E-Commerce was generated from **travel sites**.

travel site 旅游网站

E-Commerce is also expanding across the Middle East. Having recorded the world's fastest growth in the Internet usage between 2000 and 2009, the region is now home to more than 60 million Internet users. Retail, travel and gaming are the region's top E-Commerce segments, in spite of difficulties such as the lack of region-wide legal frameworks and logistical problems in cross-border transportation. E-Commerce has become an important tool for small and large businesses worldwide, not only to sell to customers, but also to engage them.

In 2012, E-Commerce sales topped $1 trillion for the first time in history.

.End.

Key Words

innovative adj.革新的，创新的
unprecedented adj.空前的，无前例的
permeation n.渗入，透过
retailer n.零售商
notification n.通知
transform v.改变
boundary n.分界线
prosperous adj.繁荣的，兴旺的
infrastructure n.基础结构
registration n.注册
EDI(Electronic Data Interchange) 电子数据交换
BTB (Business-to-Business) 企业对企业
BTC (Business-to-Consumer) 企业对消费者

参考译文 技能1 电子商务

21世纪，信息技术的快速发展和信息交换的快速增长已经为整个社会带来了新的动力和创新思想。社会对信息技术的广泛应用已经为其带来巨大的变化。这些变化不单单是在数据处理和计算上，还包括我们如何进行交流，如何安排我们的日常活动，如何教育下一代，如何做生意等。信息技术的快速发展和广泛应用以及计算机网络和因特网已经改变了许多商业的运作方式，同时也带来了前所未有的商机。现在的企业能够跨越地域的界限，穿越时空的差别，以很高的效率进行交易。电子商务已经成为本世纪市场的发展趋势。

什么是电子商务？

电子商务就是通过电子媒介来做生意。它意味着利用简单、快捷而且低廉的电子通信手段进行交易，而交易的双方不必面对面地会晤。现在，电子商务主要通过因特网和EDI进行交易。电子商务最早起源于1960年。随着计算机的广泛应用，因特网的成熟和广泛采用，信用卡的介入，安全交易协议的建立和政府的支持与推动，电子商务的发展前景越来越广阔，人们开始利用电子商务做生意。

电子商务的类型

◇ 企业内的电子网络：人们通过企业内联网交换和处理内部商业信息。

◇ 企业对企业的电子商务（BTB或B2B）：在电子商务的许多种类中，这种通过互联网和EDI进行交易的方式是最受关注的。据弗雷斯特研究公司的预测，BTB电子商务将以3倍于BTC电子商务的速度增长，因此有巨大的发展潜力。

◇ 企业对消费者的电子商务（BTC或B2C）：企业通过互联网为消费者提供在线购物，使消费者可以在线购买商品及付账。这种方式节省了零售商和消费者的时间。

◇ 消费者与消费者之间的电子商务：消费者可以通过代理商的网站在线发布自己的产品以供其他消费者竞买。

◇ 政府与企业之间的电子商务：这种交易方式是指政府通过因特网等电子媒介购买商品和服务，例如电子投标系统（ETS）。这个系统提供了在线服务的基础构架，如供应商的注册、投标通知、投标文件的下载、投标咨询、投标计划的递交和投标结果的发布等。

全球趋势

2010年，按人均花费测算，英国有当时世界上最大的电子商务市场。而欧洲国家捷克共和国的电子商务收入在其企业总收入中占据的比例最大。该国近四分之一（24%）的总营业额是通过在线渠道产生的。

在新兴经济体中，中国的电子商务存在份额在不断扩大。2009年，中国有3.84亿互联网用户，中国的网上购物销售额攀升到366亿美元。这一具大增长的背后是网购者对网购信任等级的提升。中国零售企业已经让消费者的网上购物越来越舒适。中国的跨国电子商务也迅速成长。2012年中国和其他国家之间的电子商务交易比重增至32%，达到2.3万亿元（3758亿美元），占中国国际贸易总额的9.6%。

其他金砖四国也见证了电子商务的加速增长。以2010年的价值估算，到2015年俄罗斯的电子商务市场总值预计将介于6900亿卢布（230亿美元）到9000亿卢布（300亿美元）之间，这相当于俄罗斯零售业总额的5%。长期来看，到2020年，俄罗斯的电子商务市场规模可达500亿美元。2014年，巴西的电子商务零售销售额预计将以两位数的速度快速增长，到2016年，电子市场分析人员预计巴西的电子商务零售销售额将达到173亿美元。

另一方面，印度电子商务的增长一直缓慢，尽管印度在发展电子商务方面颇具潜力，比如其经济蓬勃发展，互联网普及率增长迅速，英语语言水平高，并且拥有12亿人的庞大的消费者市场（尽管只有5000万人使用个人计算机上网，最活跃的电子商务用户只有200万到300万人）。根据公布的报告来看，从2011年到2012年电子商务的业务量增长了约50%，从2610万增长到3750万。在2012年，大约有140亿美元是从旅游网站上生成的。

在中东，电子商务也在不断扩大。2000年到2009年间，该地区的互联网使用率是世界上增长最快的，现已拥有超过6000万的网络用户。尽管还有很多困难，如跨境运输区域范围内的法律框架和后勤问题，但零售、旅游和游戏仍是该地区最大的电子商务收入组成部分。电子商务已成为全球小型和大型企业的一个重要工具，不仅向客户推销商品，也将客户纳入其中。

2012年，电子商务销售额在历史上第一次突破1000亿美元。

Skill Two Online shopping[①]

Online shopping or online retailing (Pic8.1) is a form of electronic commerce which allows consumers to directly buy goods or services from a seller over the Internet using a web browser. Alternative names are: e-web-store, e-shop, e-store, Internet shop, web-shop, web-store, online store, online storefront and virtual store. Mobile commerce (or m-commerce) describes purchasing from an online retailer's mobile optimized online site or app.

online shopping 网上购物

① 资料来源于http://en.wikipedia.org/wiki/Input/output
http://www.eshoppingindia.com/info/online-shopping-advantages.html

Pic 8.1 Online shopping

The major advantage of online shopping is the convenience it offers. By sitting back at home you can now shop anything from candles to vehicles by several clicks of mouse buttons. One of the important disadvantages of online shopping is lack of personal interaction. Another disadvantage of online shopping is tangibility factor. Seeing the picture of a product is far inferior to that of seeing it in real world. When you go for real world shopping, you can actually touch, feel or sense it with different means, but for online shopping you can only view the electronic catalogues. Even though this problem has been rectified to certain extent by use of 3D product catalogues, some online malls still use the old fashioned images in product catalogues.

Paying by **credit card** is the widely accepted method of payment for online shopping. However the other methods, like using e-checks, **PayPal & bank transfer** are also common. The method of payment is decided upon the mutual trust and familiarity between online merchant and the customer.

credit card 信用卡
PayPal 贝宝
bank transfer 银行汇款，银行转账

Retail success is no longer all about physical stores, this is evident because of the increase in retailers now offering online store interfaces for consumers. With the growth of online shopping, comes a wealth of new market footprint coverage opportunities for stores that can appropriately cater to offshore market demands and service requirements.

Online stores must describe products for sale with text, photos, and multimedia files, whereas in a physical retail store, the actual product and the manufacturer's packaging will be available for direct inspection (which might involve a test drive, fitting, or other experimentation).

Some online stores provide or link to supplemental product

information, such as instructions, safety procedures, demonstrations, or manufacturer specifications. Some provide background information, advice, or how-to guides designed to help consumers decide which product to buy.

Online shopping is a different experience and you can make the shopping creative over the Internet as you get used to it. There can be a lot of apprehensions about online shopping when you get into it for the first time. As you experience more and more of it those apprehensions get disappeared slowly. Remember that if you stick to the basics, online shopping will become more enjoyable and easier than real-world shopping.

.End.

online shopping 网上购物
credit card 信用卡
bank transfer银行汇款，银行转账
retailer n.零售商，零售店
PayPal贝宝(全球最大的在线支付平台)

参考译文 技能2 网上购物

网上购物或网上零售（图8.1）是电子商务的一种形式，它可以让消费者通过网络浏览器直接从卖家那里购买商品或服务。可用的名字有：电子网络商店、电子购物、电子商店、因特网购物、网络购物、网络商店、在线商店、线上店面和虚拟商店。移动电子商务一词描述了这种从在线零售商的移动优化网站购买物品的形式。

网上购物最大的优点是方便。你坐在家里通过点击几下鼠标就可以购买从蜡烛到车在内的所有商品。在线购物的一个最大的缺点是缺乏人际交流，另一个缺点是无法真实触碰到商品。单看产品的图片远不及在实体店看实实在在的产品。当你去实体店的时候，你能通过触摸、感觉等多种不同的方法了解商品，但是在线购物时你只能浏览电子目录。虽然这个问题已经通过使用3D电子产品目录得到改进，但是还有很多商家在产品目录中依然使用过时的图片。

信用卡支付是网上购物应用最广泛的一种支付方式。别的支付办法，像电子支票、贝宝和银行转账也是常见的方法。支付方法的选择是建立在网上商家和客户之间的相互信任和熟悉度基础之上的。

零售业的成功不再只看实体店，因为现在已经有越来越多的零售商为用户提供在线商店。网上购物的成长，对于商家来说，是丰富其商品覆盖率的机会，可以适当地迎合海外市场的需求和服务要求。

网上商店必须使用文字、图片和相应的多媒体文件去描述产品，而在实体零售店，真实的商品和生产商的包装更方便直接检查（包括实际检查，试用，查看是否合适及其他检验方法）。

一些网上商店会提供或者链接产品补充信息，如使用说明书、安全规则、产品展示或制造商规格等。一些网店还会提供相关背景信息链接、建议或指南来帮助消费者决定购买哪种商品。

网上购物是一种不同的体验，习惯之后你能通过网络使购物变得很灵活。第一次使用网络购物时你可能会有一些担心，但当你越来越多地使用它时，这种担心就会慢慢消失。请记住，如果你遵守基本规划，网上购物要比实体店购物有趣、简单得多。

Fast Reading One Introduction to Programming Languages

Computers cannot function without programs, which give them instructions. People specialized in writing programs are known as computer programmers. They construct programs by using programming languages.

Programming language, in computer science, is the artificial language used to write a sequence of instructions that can be run by a computer. These are not natural languages, such as English, Chinese, but are specified sets of words, phrases and symbols called codes, which can be combined in certain very restricted ways to instruct the computer.

However, natural languages are not suited for programming computers because they are ambiguous, meaning that their vocabulary and grammatical structure may be interpreted in multiple ways. The languages used to program computers must have simple logical structures, and the rules for their grammar, spelling and punctuation must be precise.

Programming languages date back almost to the invention of the digital computer in the 1940s. Computer languages have undergone dramatic evolution since the first electronic computers were built to assist in telemetry calculations during World War Ⅱ.

Early on, programmers worked with the most primitive computer instructions: machine language. These instructions were represented by long strings of ones and zeros. The first assembly languages emerged in the late 1950s with the introduction of commercial computers. It maps machine instructions to human-readable mnemonics, such as ADD and MOV.

The first procedural languages were developed in the late 1950s to early 1960s. FORTRAN created by John Backus and then COBOL created by Grace Hopper. The first functional language was LISP, written by John McCarthy IN THE LATE 1950s. Although heavily updated, all three languages are still widely used today. In the late 1960s, the first object-oriented languages, such as SIMULA, emerged. During the 1970s, procedural languages continued to develop with ALGOL, BASIC, PASCAL and C.

Programming languages use specific types of statements, or instructions, to provide functional structure to the program. A statement in a program is a basic sentence that expresses a simple idea—its purpose is to give the computer a basic instruction. Statements define the types of data allowed, how data is to be manipulated, and the ways that procedures and functions work.

Variables can be assigned different values within the program. The properties variables can have are called types. In many programming languages, a key data type is a pointer. Pointers themselves do not have values; instead, they have information that the computer can use to locate some other variable—that is, they point to another variable.

An expression is a piece of a statement that describes a series of computations to be performed on some of the program's variable, such as X + Y/Z, in which the variables are X, Y, and Z. The computations are addition and division. An assignment statement assigns a variable a value derived from some expression, while conditional statements specify expressions to be tested and then used to select which other statements should be executed next.

Procedure and function statements define certain blocks of code as procedures or functions that can then be returned to later in the program. These statements also define the kinds of variables and parameters the programmer can choose and the type of value that the code will return when an expression accesses the procedure or function.

The most commonly used programming languages are highly portable and can be used to effectively solve diverse types of computing problems. Programming languages can be classified as either low-level programming languages or high-level programming languages.

Low-level programming languages, for example, assembly languages, are the most basic type of programming languages. Assembly languages are very close to machine languages, but they must still be translated into machine language. How to understand machine languages? Machine languages are a sequence of 1s and 0s, called bits, and can be understood directly by a computer.

In the same way, high-level programming languages are close to human natural languages, and also must first be translated into machine languages by a compiler before they can be understood and processed by a computer. Examples of high-level programming languages are FORTRAN, ALGOL, Delphi, SNOBOL, Pascal, C, C++, Visual C++, Visual C #.NET, COBOL, BASIC, LISP, PROLOG, Visual Basic, Visual FoxPro, Java, and Ada.

For this reason, programs written in a high-level programming language may take longer to execute and use up more memory than programs written in an assembly language. High-level programming languages are more similar to normal human languages than low-level programming languages. Therefore, it is easier for the programmer to write larger and complicated programs faster.

.End.

参考译文 编程语言介绍

如果没有程序给予计算机指令，计算机是无法运行的。专门编写程序的人被称为计算机程序员。他们用编程语言来构建程序。

在计算机科学中，编程语言是用来编写可被计算机运行的一系列指令的人工语言。这种语言与英语、汉语等自然语言不同，它是被称为代码的专门词汇、短语及符号的集合，用特定的方式把这些代码组合起来指导计算机运行。

无论如何，自然语言都不适合计算机编程，因为它们能引起歧义，也就是说它们的词汇和语法结构可以用多种方式进行解释。用于计算机编程的语言必须具有简单的逻辑结构，而且它们的语法、拼写和标点符号的规则必须精确。

编程语言几乎可以追溯到20世纪40年代数字计算机发明之时。自从第一代电子计算机在第二次世界大战中用于遥测计算以来，计算机语言已发生了巨大的变化。

早期程序员使用最原始的计算机指令——机器语言来工作。这些指令由一长串的0、1代码组成。最早的汇编语言，随着商业计算机的推出，出现于20世纪50年代末。汇编语言能将机器指令转换成易读、易管理的助记符，如ADD、MOV等。

最早的过程语言是在20世纪50年代末到20世纪60年代初开发的。约翰·巴克斯创造了FORTRAN语言，之后格雷斯·霍珀创造了COBOL语言。第一种函数式语言是LISP，由约翰·麦卡锡编写于20世纪50年代末。这三种语言今天仍被广泛使用，但经历过大量修改。20世纪60年代末，出现了最早的面向对象语言，如SIMULA语言。在20世纪70年代，过程语言继续发展，出现了ALGOL、BASIC、PASCAL和C语言等。

编程语言使用特定类型的语句或指令来给程序提供功能结构。程序中的一条语句表达一个意思简单的基本句子，其目的是给计算机提供一条基本指令。语句对允许的数据类型、数据如何处理以及过程和函数的工作方式进行定义。

变量在程序中可以被赋予不同的值。变量具有的属性被称作类型，在许多编程语言中，一种关键的数据类型是指针。指针本身没有值；但是，它们含有计算机可以用来查找某个其他变量的信息——也就是说，它们指向另一个变量。

表达式是语句的一段，描述要对一些程序变量执行的一系列运算，如X+Y/Z，其中X、Y和Z为变量，运算方法为加和除。赋值语句给一个变量赋予来自某个表达式的值，而条件语句则是指定要被测试、然后选择接下来应该执行哪个语句的表达式。

过程与函数语句将某些代码块定义为以后可在程序中返回的进程或函数。这些语句也规定了程序员可以选择的变量与参数种类，以及当一个表达式使用过程或函数时代码将返回的值的类型。

最常用的编程语言具有很高的可移植性，可以用于有效地解决不同类型的计算问题。编程语言可划分为低级编程语言和高级编程语言。

低级编程语言，比如汇编语言，是编程语言中最基础的类型。汇编语言非常接近机器语言，但仍然得翻译成机器语言。如何理解机器语言呢？机器语言是计算机能够直接识别的被称为比特的1、0代码的序列。

同样地，高级编程语言接近于人类自然语言，在计算机能够理解和处理之前也必须首先被翻译成机器语言。FORTRAN、ALGOL、Delphi、SNOBOL、Pascal、C、C++、Visual C++、Visual C#.NET、COBOL、BASIC、LISP、PROLOG、Visual Basic、Visual FoxPro、Java和Ada都是高级编程语言。

由于上述原因，与用汇编语言编写的程序比较起来，用高级编程语言编写的程序可能运行的时间更长，占用的内存更多。高级编程语言比低级编程语言更类似于正常的人类语言，因此，程序员容易更快地编写更庞大和更复杂的程序。

Fast Reading Two Safe Shopping Online

Privacy of personal information is a significant issue for some consumers. Many consumers wish to avoid spam and telemarketing which could result from supplying contact information to an online merchant. In response, many merchants promise not to use consumer information for these purposes.

Many websites keep track of consumer shopping habits in order to suggest items and other websites to view. Brick-and-mortar stores also collect consumer information. Some ask for a shopper's address and phone number at checkout, though consumers may refuse to provide it. Many larger stores use the address information encoded on consumers' credit cards (often without their knowledge) to add them to a catalog mailing list. This information is obviously not accessible to the merchant when paying in cash.

Many online malls use cookies to track the user activity for showing relevant results to maximize the shopping experience. It can also be used to track your personal details. Thus before you pass your personal information, ensure the credibility of the online merchant. Good companies post their privacy policy (that is how they are going to use the personal information about you) on the website. Read carefully!

The important considerations after privacy are the security and safety features used by online malls. Remember that good websites are made in compliance with industrial standards such as SSL (secured socket layer). These standards use encryption technology to transfer information from your computer to online merchant's server.

By using SSL, the information you sent is scrambled, this means it is not possible to get details without encryption code. Since this is done automatically in merchants server, it can be ensured that your personal details are secure. Thus make sure that you always do payment over SSL.

When the company you want to deal is new to you, try to get maximum information about them before making any orders. Keep your password secret and make it in such a way that others may not be in a position to guess it.

Credit card transactions are considered to be the safest mode of payment for online shopping. Make yourself understood with the company's policies especially on how they are going to keep your financial and personal data secured. Keep printed copies of purchase order and confirmation details, so that it can be used in the event of disputes.

.End.

参考译文 安全的网上购物

对于一些消费者来说，个人信息保密是一个重要的问题。许多消费者都希望摆脱因向网上商家提供联系信息而出现的垃圾邮件和电话推销。作为回应，许多商家承诺不会出于这些目的来使用消费者信息。

许多网站会保留消费者的购物习惯，以便向其推荐其他商品或网站。实体店也会收集消费者信息，一些商家要求顾客在付款时提供地址和电话号码，但消费者可以拒绝提供。许多大商店会使用消费者信用卡上的地址信息编码(通常在他们不知情的情况下)，并将它们添加到邮寄目录中，这些信息显然不是向商家付现金时留下的。

许多网上商城使用cookie跟踪用户活动，以显示相关搜索结果来优化购物体验。它也可以用来记录你的个人信息，因此，在你留下个人信息之前，要确保网上商家的可信度。好的公司会在网站上发布其隐私政策(即它们将如何使用你的个人信息)，仔细阅读这些条款。

排在隐私之后的关于网上商城的主要考量方面是安全性和安全保护措施。记住，好的网站是符合工业标准如SSL(安全套接字层)的。这些标准使用加密技术将你电脑中的信息传输到在线商家的服务器上。

使用SSL发送信息，信息会经过置乱处理，这意味着信息不可能被获取，也就是说，没有加密代码是无法了解详细信息的。因为这是由商家服务器自动完成的，所以可以保证你的个人信息是安全的。因此确保你总是通过SSL付款。

如果你要和新店家打交道，在做出任何订单前，一定要最大限度地获取它们的信息。设置你的密码，确保你的密码别人不能轻易猜到。

网上购物中信用卡交易被认为是最安全的付款方式。要清楚交易公司的政策，尤其是在如何保证你的金融和个人信息安全方面的政策。保留打印的采购订单副本和确认的细节凭证，以便在发生纠纷时使用。

Exercise

Ex 1 What is online shopping ? Try to describe it in some words from the passage.

Ex 2 How many types are there in the E-Commerce? Describe them in some words.

Ex 3 Fill in the table below by matching the corresponding Chinese or English equivalents.

	零售商
online	
	银行汇款
BTB	
	电子数据交换
E-Commerce	
	注册

Part B Practical Learning

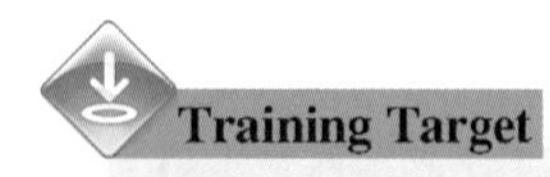

Training Target

In this part, our target is to train students how to use specialized English knowledge to finish professional tasks in English environment. It can achieve the purpose of using professional English through the tasks.

Task One Discuss the Differences Between Buyer and Seller

The important things for seller:

First: Online form should attract customers.

Secondly: The quality of the products should be good.

Thirdly: Ensure customer information security.

Fourthly: Right service attitude is irreplaceable.

The important things for buyer:

First: Find the product which you want to buy

When looking to purchase anything online, starting by checking if the stores you usually shop have an online store. If you can't find a familiar name, don't hesitate to ask; you can find many helpful people online in chat rooms, forums, or mailing lists. You can ask for references or see if a site has any customer comments. You can also check the Better Business Bureau to see if there have been any complaints against a company.

Secondly: Read everything about the products carefully

One of the most important things is to read all the information you can find about the specific product you are ordering. The last thing you want to do is receiving an item with a wrong size or a wrong color etc. Do not rely solely on pictures provided. If the online store does not have enough information about a product, you can try to E-mail or call to see if they can clarify for you.

Thirdly: Payment options

Check to see what kind of payment a site accepts. The most common way to buy things online is to use a credit card, which is very safe these days as long as a site has a secure connection. Some sites will let you pay by check. With this option, you usually place your order and then mail the payment, and the store will hold onto your order until the payment is received.

Fourthly: Shipping cost

One of the deciding factors for you when it comes to ordering craft supplies online is the shipping cost. In some cases, the shipping charge may be more than the cost of your purchase; again, read everything carefully! But sometimes people bite the bullet simply for the convenience of getting items shipped directly to their homes.

Last but not least: How about returns?

Make sure you are clear about the site's return policy. Some places have very strict policies and also charge restocking fees. You should be aware of what you are up against in case you need to return your products.

Task Two Students Trade Online

After finishing the online store design and knowing the principles of online trade, in this task, students will trade online. Some students should act as the seller, other students should act as the buyer. Then they can trade online.

Part C Occupation English

Training Target

In this part, students are supposed to practice the dialogue, who act as staff of service center. Offering help to customers in operation system installations is one of the common tasks in their future job.

Trouble Shooting in Office Program 解决办公软件问题

Post : Representative of Microsoft Technical Support（岗位：微软技术支持人员）

微软技术支持人员要非常熟悉常用软件的操作，特别是视窗操作系统和办公系列软件。

A : Microsoft Technical Support. Can I help you?

微软技术支持。我能为您做些什么？

B : Hello, morning! I have a problem in Microsoft Word 2003. Something seems wrong when I edit a document. That is, there exists a line on top of every page.

上午好。我在使用微软Word 2003方面需要你的帮助。我在编辑文档的时候总是出错。每页最上面都有一条横线。

A : Well, if there is a horizontal line on top of every page, it is probably because you have deleted the header of your document without removing the underline.

哦，如果每页最上方都有一条水平的线，那可能是因为您删除了文档的页眉，而下划线还留在那里。

B : Exactly. I deleted its header. What can I do to remove the line?

就是这样。我删除了页眉。我怎么做才能把这条线也移走呢？

A : That’s not difficult. You can remove the line by editing the border of its Header text.

这个不难。您可以通过编辑“页眉文本”的边界来删除这条横线。

B : Please, I need the details.

请你说得更详细一些。

A : OK, first, in a Word page, go to the Edit menu, and choose Header and Footer. Put the cursor in the header area, and press Control and A at the same time. That will select everything in the area. Then go to the Format menu, and select Borders and Shading. Go to the border section and remove the underline. 好的。首先，在一个Word页面中，打开“编辑”菜单，选择“页眉和页脚”。将光标放在页眉处，同时按下“Ctrl”和“A”键。这样就会选择该区域内的所有文字和符号。再打开“格式”菜单，选择“边框和阴影”。打开“边框”部分，就可以删除这条下划线了。

B : It is done. Thank you so much!

我已经把它删掉了。非常感谢。

A : Anytime. Just call us if you need any further help.

愿意随时为您效劳。如果您还需要帮助，请随时拨打电话联系我们。

Word Building

前缀/后缀由一个或几个字母组成，放在词根或单词之前/之后，组成一个新词。

(1) auto-（前缀）：自动的

alarm：报警器 ———— autoalarm： 自动报警器

code：编码 ———— autocode：自动编码

(2) -ee（后缀）：被……的人，受……的人

employ：雇用 ———— employee：雇员

test：测试 ———— testee：被测验者

(3)-th（后缀）：动作，性质，过程，状态

true：真实的，真正的 ———— truth：事实，真理

weal：福利，幸福 ———— wealth：财富

(4)-al（后缀）：属于……的，与……有关的

digit：数字 ———— digital： 数字的

Ex Translate the following words and try your best to guess the meaning of the word on the right according to the clues given on the left.

train	训练（动词）	trainee	________
pay	薪水（名词）	payee	________
grow	生长（动词）	growth	________
deep	深的（形容词）	depth	________
education	教育（名词）	educational	________
nature	自然（名词）	natural	________
rotation	旋转（名词）	autoratation	________
biography	传记（名词）	autobiography	________

Exercise

Ex 1 What is E-Commerce ? Try to describe it in some words from the passage.

Ex 2 Fill in the table below by matching the corresponding Chinese or English equivalents.

	电子商务
BTB	
	网上购物
BTC	
	贝宝
restricting access	
	信用卡

Ex 3 Choose the best answer to the following questions according to the text we learnt.

1. E-Commerce is doing business through ________.
   A. interfaces　　B. programs
   C. kernel techniques　　D. electronic media
2. Which is not the type of E-Commerce ?
   A. BTB　　B. BTC
   C. ETS　　D. Internet Explore
3. Which is the widely accepted method of payment for online shopping?
   A. credit card
   B. e-checks
   C. PayPal
   D. bank transfer
4. Many online malls use ________ to track the user activity for showing relevant results to maximize shopping experience.
   A. cookies
   B. Windows NT/2000
   C. Windows 98
   D. credibility
5. Good companies post their ________ on the website.
   A. privacy policy
   B. multi-user operating
   C. IE 6.0
   D.A and C

Reference

[1] 金志权.计算机专业英语教程[M].4版.北京：电子工业出版社，2008.

[2] 白建忠.计算机专业英语[M].2版.北京：中国水利水电出版社，2007.

[3] 支丽平.计算机专业英语[M].北京：中国水利水电出版社，2010.

[4] Timothy J O’Leary ,Linda I O’Leary.计算机专业英语- Computing Essentials[M]. 北京：高等教育出版社，2008.

[5] http://familycrafts.about.com/od/beforeyoubuy/bb/suppliesonline.htm.

[6] http://en.wikipedia.org/wiki/Input/output.

[7] http://www.eshoppingindia.com/info/online-shopping-advantages.html.

[8] http://www.computerhope.com/jargon/r/router.htm.

[9] http://www.answers.com/topic/network-switch.